Lithium-ion Batteries
Materials and Applications

Edited by

Inamuddin[1,2,3], Rajender Boddula[4], Mohammad Faraz Ahmer[5] and Abdullah M. Asiri[1,2]

[1]Chemistry Department, Faculty of Science, King Abdulaziz University, Jeddah 21589, Saudi Arabia

[2]Centre of Excellence for Advanced Materials Research, King Abdulaziz University, Jeddah 21589, Saudi Arabia

[3]Department of Applied Chemistry, Faculty of Engineering and Technology, Aligarh Muslim University, Aligarh-202 002, India

[4]CAS Key Laboratory of Nanosystem and Hierarchical Fabrication, National Center for Nanoscience and Technology, Beijing 100190, PR China

[5]Department of Electrical Engineering, Mewat College of Engineering and Technology, Mewat-122103, India

Published by **Materials Research Forum LLC**
Millersville, PA 17551, USA

Published as part of the book series
Materials Research Foundations
Volume 80 (2020)
ISSN 2471-8890 (Print)
ISSN 2471-8904 (Online)

Print ISBN 978-1-64490-090-1
eBook ISBN 978-1-64490-091-8

Distributed worldwide by

Materials Research Forum LLC
105 Springdale Lane
Millersville, PA 17551
USA
https://www.mrforum.com

Manufactured in the United States of America
10 9 8 7 6 5 4 3 2 1

Table of Contents

Preface

Lithium-ion batteries (LIBs) are one of the most important energy storage systems to cope with the energy cricis challenges faced in electronics and automobile industries. Lithium-ion batteries exhibit high voltage window, high energy density, low-rate discharge, and long life span. Lithium-ion batteries are preferred to use in consumer applications such as phone batteries, laptops, and hybrid electric vehicles, etc. However, the development of lithium-ion batteries with high power density is challenging. Therefore, research efforts devoted to designing high-performance electrode materials and electrolytes for lithium-ion battery applications. The research in the area of lithium-ion batteries has been in progress towards the development of practically viable technologies. The lithium-ion batteries have an incredible future, but still more research and development studies are needed to commercialize them at a large performance scale.

The book on "Lithium-ion Batteries: Materials and Applications" provides a comprehensive overview of lithium-ion batteries and their applications in consumer and industrial sectors, including safety and reliability features. This book introduces systematic and broad coverage of the properties, theoretical insights, and experimental pieces of evidence from laboratory scale to commercial scale applications of lithium-ion batteries. The book chapters summarize the mechanisms, principles, analytical insights, fundamental concepts and recent advances of anodes, cathodes, electrolytes, separators, binders, fabrication of device assemblies, electrochemical performance, etc. This book delivers information on basic research, case studies, and lithium-ion batteries applications. It is a well structured and essential resource for scientists, undergraduate and postgraduate students, faculty members, R&D professionals, energy chemists, and industrial experts.

Inamuddin[1,2,3], Rajender Boddula[4], Mohammad Faraz Ahmer[5] and Abdullah M. Asiri[1,2]

[1]Chemistry Department, Faculty of Science, King Abdulaziz University, Jeddah 21589, Saudi Arabia

[2]Centre of Excellence for Advanced Materials Research, King Abdulaziz University, Jeddah 21589, Saudi Arabia

[3]Department of Applied Chemistry, Faculty of Engineering and Technology, Aligarh Muslim University, Aligarh-202 002, India

[4]CAS Key Laboratory of Nanosystem and Hierarchical Fabrication, National Center for Nanoscience and Technology, Beijing 100190, PR China

[5]Department of Electrical Engineering, Mewat College of Engineering and Technology, Mewat-122103, India

Lithium-ion Batteries - Materials and Applications
Materials Research Foundations **80** (2020) 1-27

Materials Research Forum LLC
https://doi.org/10.21741/9781644900918-1

Chapter 1

Fabrication of TiO$_2$ Materials for Lithium-ion Batteries

Chang-Seop Lee [1]*, Hasan Jamal[1]

[1] Department of Chemistry, Keimyung University, Daegu 42601, Korea

* surfkm@kmu.ac.kr

Abstract

Among transition metal oxides, TiO$_2$ based materials are quite promising and extensively studied in lithium-ion batteries (LIBs) owing to their high safety, low cost, and low volume expansion (< 4%). In the present study, nanocomposites of TiO$_2$ (nanoparticles, nanorods, nanofibers) with graphene (TiO$_2$@rGO) were successfully synthesized by hydrothermal and calcination treatment. Coaxial SnO$_2$@TiO$_2$ nanotube arrays (SnO$_2$@TNTs) were assembled through electrochemical fabrication technique using pristine TiO$_2$ nanotube arrays (TNTs) and metallic-Sn. All as-synthesized products were applied as negative materials for LIBs. Their physicochemical properties were investigated via scanning and transmission electron microscope (SEM/TEM), X-ray diffraction (XRD), energy-dispersive X-ray spectroscopy (EDX), Raman spectroscopy and Brunauer–Emmett–Teller (BET). Electrochemical testing was performed with galvanostatic charge/discharge system. TiO$_2$ nanoparticles nanocomposites with rGO exhibited rate capacity of 155 mAhg^{-1} at C-rate of 0.5 C. After applying a high C-rate of 20 C, its rate capability was 109 mAhg^{-1} at 0.5 C, showing capacity loss of 30 %. Reversible capacity delivered by SnO$_2$@TNTs was 113 µA h cm^{-2} at current density of 100 µA cm^{-2} after 50 cycles, higher than that by pristine TNTs (51.6 µA h cm^{-2}).

Keywords

TiO$_2$ Nanostructure, Graphene, Lithium-Ion Batteries, TiO$_2$ Nanotube Arrays, Hydrothermal, Electrochemical Fabrication

Contents

Materials Research Forum LLC
https://doi.org/10.21741/9781644900918-1

Prefix	Symbol for Prefix
TiO_2 nanoparticles	TiO_2 NPAS
TiO_2 nanorods	TiO_2 NRDS
TiO_2 nanofibers	TiO_2 NFBS
TiO_2 nanoparticles/rGO	TiO_2 NPAS@rGO
TiO_2 nanorods/rGO	TiO_2 NRDS@rGO
TiO_2 nanofibers/rGO	TiO_2 NFBS@rGO

1. Introduction

With increasing development of portable electronics, solar energy, wind energy, and hybrid/full electric vehicles (EVs), stationary energy storage devices are stringently needed [1,2]. Rechargeable LIBs are more advanced energy storage systems than other battery technologies due to their higher volumetric and gravimetric energy density. Hybrid vehicles account for 3% of all cars today. However, it has been predicted that their number will increase rapidly in the coming years [3]. Utilizing EVs instead of traditional combustion engine can significantly reduce pollution. Thus, LIBs are anticipated to have a very positive effect on sustainable-energy economy and the environment [4,5].

The anode is the key part in LIBs. It performs a critical role in determining the capacity of the device. According to previous studies [6,7], an ideal anode should meet the following conditions: (i) exposed large surface area offering more channels for Li-insertion, (ii) low volume expansion for lithiation and de-lithiation essential for stable cycling, (iii) low intrinsic resistance which supports fast lithiation and de-lithiation, (iv) large pore size and shorter path length that favor the fast diffusion of Li-ions which is paramount for superior rate capability, (v) low lithium intercalation potential, (vi) cheap price, and (vii) environmental benignity. Based on the mechanism of lithium storage, negative material for LIBs can be classifieds into the succeeding classes: (i) carbon related materials like graphite, graphene, amorphous carbon, and carbon nanotube; (ii) alloy/dealloy materials such as Si, Al, Bi, Sn, and Ge; (iii) transition metal oxides (M_xO_y, M = Ni, Co, Cu, Mn, Fe, etc.); and 4) metal nitrides, metal sulphides, and metal phosphides [8–10]. Figure 1 shows the capacity of some potential anode materials and corresponding potential versus Li/Li^+. Overall, transition metal oxides have relatively higher capacity and potential than other electrode materials.

Lithium-ion Batteries - Materials and Applications   Materials Research Forum LLC
Materials Research Foundations **80** (2020) 1-27   https://doi.org/10.21741/9781644900918-1

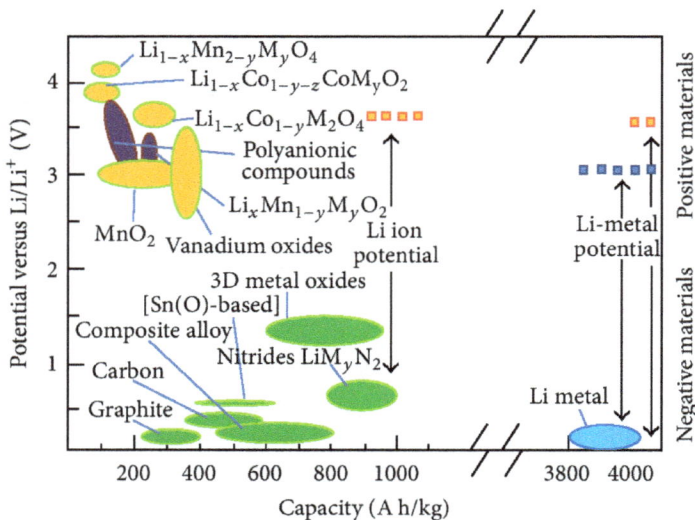

Figure 1. Comparison of some potential anode materials for lithium ion batteries [18].

Among transition metal oxides, TiO_2 has been extensively investigated as negative candidate for LIBs due to its environmental benignity, corrosion resistivity, abundant availability, high safety, and non-toxic nature. TiO_2 exhibits high discharge voltage plateau (> 1.7 V vs. Li/Li$^+$) and outstanding cycling reversibility [11,12]. In addition, TiO_2 undergoes very small volume expansion (< 4 %) at the time of lithium insertion and extraction process, prompt to longer cyclibility and resilience [13]. These characteristics make TiO_2 a favorable negative material for high energy storage [14–16].

However, TiO_2 has some limitations such as poor rate capability, low capacity, and low electrical conductivity. Reversible lithiation-delithiation in TiO_2 occurs according to the following reaction [17].

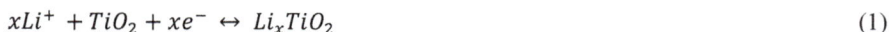

$$xLi^+ + TiO_2 + xe^- \leftrightarrow Li_xTiO_2 \tag{1}$$

Where, x ranges from 0 to 1 largely depending on the morphology, polymorph, and particle size of TiO_2. Hence, the discharge capacity of TiO_2 largely relies on structural parameters of TiO_2. Table 1 outlines structural and lithiation quantity of various TiO_2

polymorphs [18]. Of these, TiO_2 anatase, rutile, brookite, and bronze phases for LIBs have been outlined. However, TiO_2 has some major drawbacks such as low electrical conductivity (10^{-12} - 10^{-7} s cm^{-1}) and low lithium-ion diffusion coefficient (10^{-15} - 10^{-9} cm^2 s^{-1}), leading to poor rate capability and thus to the unavailability of open channels for Li-ion insertion [19–21].

Table 1. Structural and electrochemical properties of various TiO_2 polymorphs [13].

Structure	Space group	Density (g cm^{-3})	Lattice parameter values	Lithiation quantity (mole)	
				Bulk	**Nano**
Rutile	Tetragonal P4$_2$/mnm	4.13	a = 4.59, c = 2.96	0.1	0.85
Anatase	Tetragonal 14$_1$/amd	3.79	a = 3.79, c = 9.51	0.5	1.0
Brookite	Orthorhombic Pbcv	3.99	a = 9.17, b = 5.46, c = 5.14	0.1	1.0
TiO$_2$–B (bronze)	Monoclinic C2/m	3.64	a = 12.17, b = 3.74, c = 6.51, β = 107.298	0.71	1.0
TiO$_2$–II (Columbite)	Orthorhombic Pbcn	4.33	a = 4.52, b = 5.5, c = 4.94		
TiO$_2$-H (hollandite)	Tetragonal 14/m	3.46	a = 10.18, c = 2.97		
TiO$_2$-III (baddeleyite)	Monoclinic P2$_1$/c		a = 4.64, b = 4.76, c = 4.81, β = 99.28		
TiO$_2$-R (ramsdellite)	Orthorhombic Pbmn	3.87	a = 4.9, b = 9.46, c = 2.96		
TiO$_1$-O I	Orthorhombic				
TiO$_2$–O II	Orthorhombic				

Due to aforementioned shortcomings of TiO_2 anodes, several techniques have been evolved to address these issues. Four approaches have been suggested to improve the electrochemical performance of TiO_2 electrode. The first approach is to crop electronic and ion transport length by fabricating different types of TiO_2 nanostructures. The second approach is by reducing charge transfer resistance, compensating volume expansion, and improving lithium-ion diffusion coefficient by coating or engulfing with carbonaceous materials. The third approach is by restraining nearly zero volume change during the insertion of Li^+ ions by fabricating metal oxide core-shell nanocomposites with TiO_2 materials. The fourth approach focuses improving the intrinsic nature and increasing the internal surface area opens more reactive sites for Li-ion transport by doping with other atom or ions.

2.   Synthesis of TiO_2/graphene nanocomposites and metal oxides core-shells $SnO_2@TiO_2$ nanotube hybrids

2.1   Preparation of TiO_2 NRDS

Titanium sulfate (2.0 g) was first dispersed in 40 ml of deionized water (DI). Then 20 ml ammonia solution was dissolved drop-wise for 1 h in the preceding solution. The obtained mixture was transferred into autoclave, heated to 180 °C and then kept for 24 h. Afterward, the obtained products were thoroughly washed with water and dried at 60 °C for 24 h. After calcination [13], anatase phase of TiO_2 NRDS were obtained.

2.2   Synthesis of TiO_2 NFBS

Typically, 0.5 g of TiO_2 anatase powder (particle size: 25 nm) was transferred into Teflon reactor and suspended in 65 ml of 10 M sodium hydroxide solution. The Teflon reactor was sealed and installed into a stainless steel autoclave for 22 h at 190 °C. PH 7 was obtained after washing with DI water. The collected sample was dried at 60 °C for 12 h. The obtained sample powder was supplemented with sodium ions by stirring in 0.1 M hydrochloric acid and washed with DI water until pH 7 was obtained. After drying the sample in vacuum at 60 °C, the anatase phase of TiO_2 NFBS were achieved by annealing in Ar flow at 800 °C for 1h.

2.3   Synthesis of TiO_2 nanocomposites with graphene

First, an aqueous solution (200 ml, 2 mg/ml) of graphene oxide (GO) was dispersed with 1.2 g of anatase TiO_2 NPAS (nanopowder with particle size of 25 nm, Sigma-Aldrich). The obtained sample was refluxed at 90 °C for 2 h, then cooled at room temperature. After cooling, sample was centrifuged and washed with H_2O and C_2H_5OH. In obtained products, 20 ml C_2H_5OH and 10 ml distilled H_2O were stirred and autoclaved at 180 °C for 24 h in 500 ml Teflon lined autoclave. To obtain highly crystalline composites,

Materials Research Forum LLC
https://doi.org/10.21741/9781644900918-1

samples were calcinated in argon (Ar) flow at 400 °C for 4 h. Similar procedure was used to obtain nanocomposites of other TiO_2 nanostructures with graphene [15]. Schematic representation for the fabrication of TiO_2@rGO nanocomposites are displayed in Fig. 2.

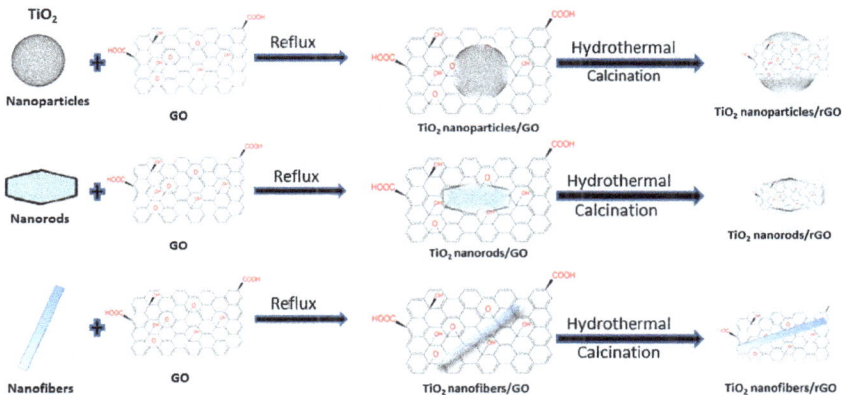

Figure 2. Synthesis procedure of TiO_2/graphene nanocomposites [13].

2.4 Synthesis of coaxial SnO_2@TiO_2 nanotube hybrids

TNTs were produced in electrolyte solution (ethylene glycol, 0.25 wt% NH_4F, 2 vol% H_2O) for 15-30 min at 50 V. To obtain crystalline form of TiO_2, sample was annealed at 450 °C for 1 h under air atmosphere. SnO_2-core layers were grown using sequential fabrication without compromising the pore structure.

For electrochemical deposition process, a conventional two-electrode system (Pt as counter electrode, annealed TNTs as working electrode) PARSTAT 2273 was used. The electrolyte consisted of 18 mM $SnCl_2.H_2O$ buffered with 50 mM $Na_3C_6H_5O_7.2H_2O$. Metallic Sn deposition relied on regulated pulse signals in microsecond span. Firstly, for short time current negative pulse was activated for the deposition of Sn. Secondly, to discharge the barrier layer capacitance, an extended positive impulse was imposed. Finally, for the conversion of metallic Sn to SnO_2, previously prepared products were annealed in air atmosphere for 90 min at 500 °C. Systematic illustration for the fabrication of SnO_2@TNTs hybrids can be found in Fig. 12.

3.   Fabrication of cell for electrochemical characterization

3.1   Electrochemical measurements for TiO_2/graphene nanocomposite

For electrochemical evaluation, electrodes were prepared with active material powder (80 wt%), carbon black as conducting agent (10 wt%), and polyvinylidene fluoride (PVDF, 10 wt%) as a binder. They were dissolved in N-methylpyrrolidone (NMP) and stirred for 30 min. The prepared slurry was coated onto copper foil using "doctor blade" technique. After evaporating the solvent at 80 °C under vacuum, coin cells (CR-2032) were fabricated under Ar (O_2 and H_2O content < 1 ppm). The electrolyte was 1 M Lithium hexafluorophosphate ($LiPF_6$) containing in ethylene carbonate (EC) dimethyl carbonate (DMC) 1:1 vol. ratio. Fabricated coin cells were charged and discharged galvanostatically (Neware Co, Ltd, Shenzhen, China) at different C-rates between 1.0 and 3.0 V.

3.2   Electrochemical tests for coaxial SnO_2@TiO_2 nanotube hybrids

Electrochemical performance was investigated by fabricating coin cells in Ar filled glove box (MBRAUN). Active material and Li-foil were used as a working and reference electrode. A 1 M $LiPF_6$ consists of (EC) and diethyl carbonate (DEC) with a volume ration of 1:1, used as a electrolyte. Celgard 2300 was used as a separator. The assembled cells were charged/discharged galvanostatically at various current rates. Impedance spectra measurements were carried out at several frequencies by implementing 10 mV AC voltage amplitude signals.

4.   Characterization of TiO_2/graphene nanocomposites

4.1 TiO_2 /graphene nanocomposites

Based on morphology, different types of TiO_2 were synthesized using titanium sulphate and commercial TiO_2 nanopowder as precursor. Nanocomposites of TiO_2/graphene were synthesized with morphologically different forms of TiO_2 nanostructures (nanoparticles, nanorods, and nanofibers) through hydrothermal and calcination approach. All nanocomposites of TiO_2 with graphene were reduced in Ar at 400 °C for 4 h.

4.1.1 SEM

SEM images of TiO_2 before synthesis of TiO_2 nanocomposites (nanoparticles, nanorods, nanofibers) with graphene are shown in Figs. 3a-3e. All TiO_2 nanostructures had distinct morphology as shown in Figs. 3a-3e. SEM images of TiO_2@rGO nanocomposites after calcination at 400 °C for 4 h under Ar are presented in Figs. 4a-4f. After calcination under Ar atmosphere, morphologies of all nanocomposites were preserved with little agglomeration (Figs. 4a-4f).

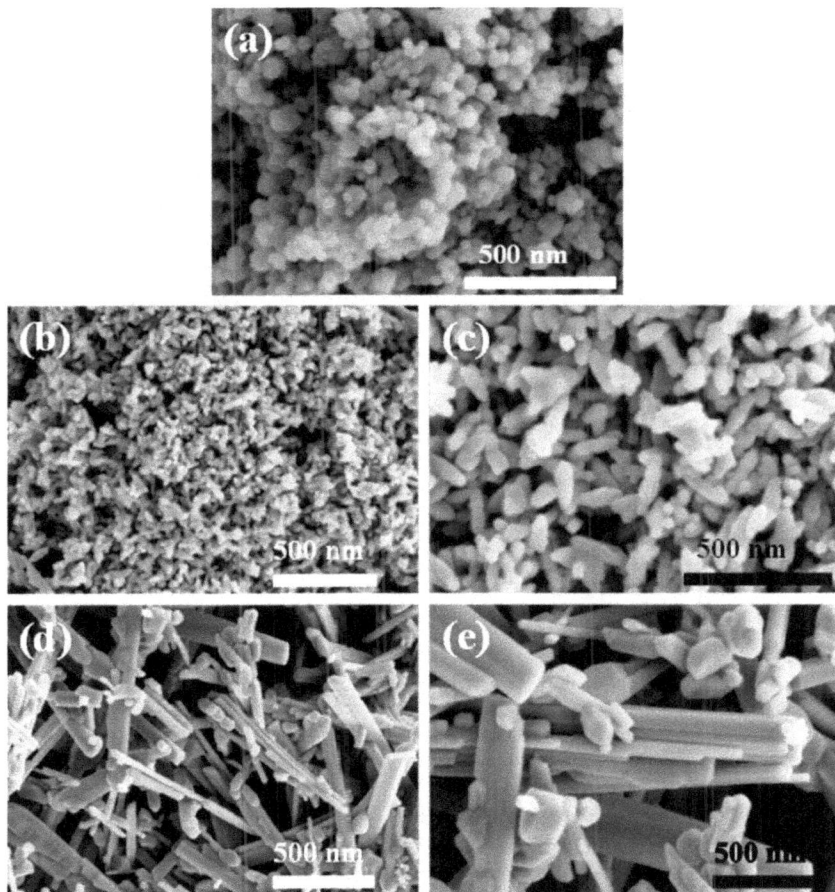

Figure 3. FE-SEM images of anatase TiO₂ before synthesis of nanocomposite with graphene. (a) TiO₂ NPAS. (b, c) TiO₂ NRDS, (d, e) TiO₂ NFBS [13].

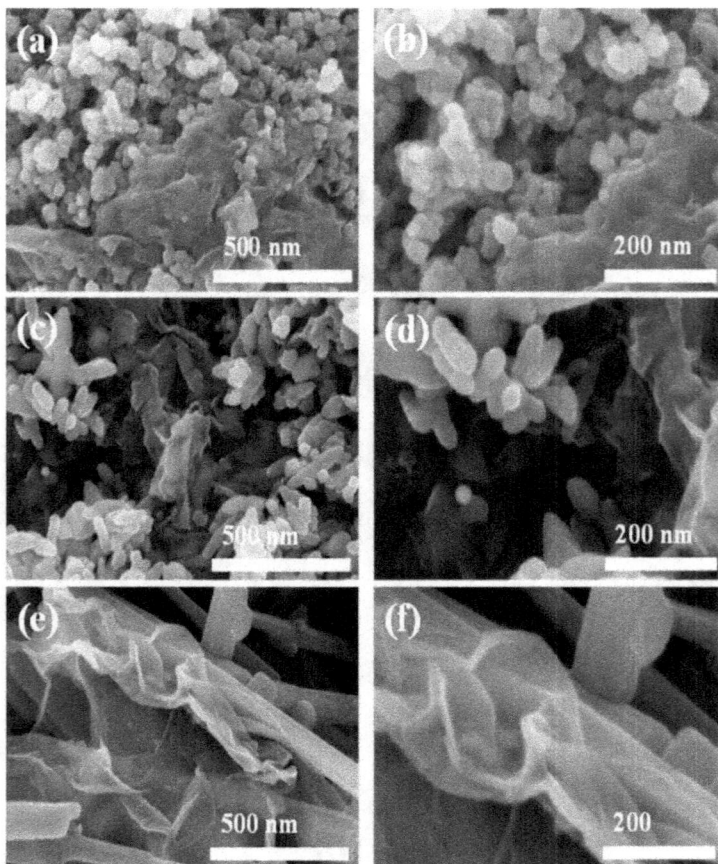

Figure 4. FE-SEM images after fabricating nanocomposite of anatase TiO$_2$ with graphene (TiO$_2$@rGO). (a, b) TiO$_2$ NPAS@rGO, (c, d) TiO$_2$ NRDS@rGO, (e, f) TiO$_2$ NFBS@rGO [13].

4.1.2 TEM

TEM images of TiO$_2$ nanostructures before fabrication of nanocomposite with graphene are depicted in Figs. 5a-5c. TiO$_2$ NPAS and TiO$_2$ NRDS had average diameters of ~30 nm and ~180 nm, respectively. Lengths of TiO$_2$ NFBS were up to several hundred

Materials Research Forum LLC
https://doi.org/10.21741/9781644900918-1

nanometers as seen in Figs. 5a-5c.TEM images of TiO_2@rGO nanocomposites after calcination at 400 °C for 4 h under Ar are illustrated in Figs. 6a-6f. High-resolution TEM images of TiO_2 nanocomposites with graphene are shown in Figs. 6b, 6d, and 6f indicated that anatase TiO_2 had a well-defined crystal structure. The d-spacing of 0.353 nm is in well agreement with (101) anatase plane of TiO_2 and that of 0.35 nm is in agreement with (002) plane of graphene. TiO_2 nanostructures were dispersed successfully among graphene sheets after hydrothermal and calcination treatment as shown in Figs. 6a-6f. The successful dispersion of TiO_2 between graphene layers provided high surface area, thus endowing more reaction sites for lithium insertion and de-insertion.

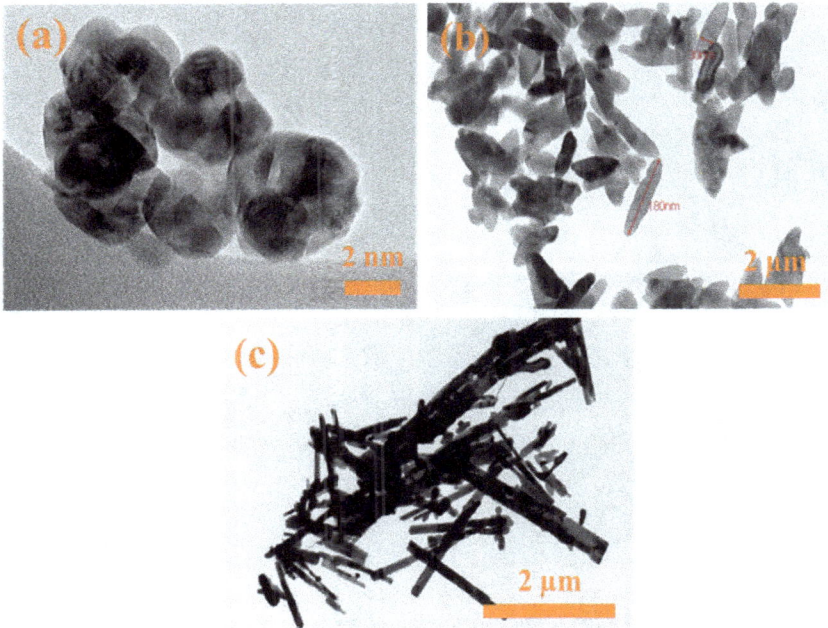

Figure 5. HR-TEM images of anatase TiO_2 before synthesis of nanocomposite with graphene. (a) TiO_2 NPAS, (b) TiO_2 NRDS, (c) TiO_2 NFBS [13].

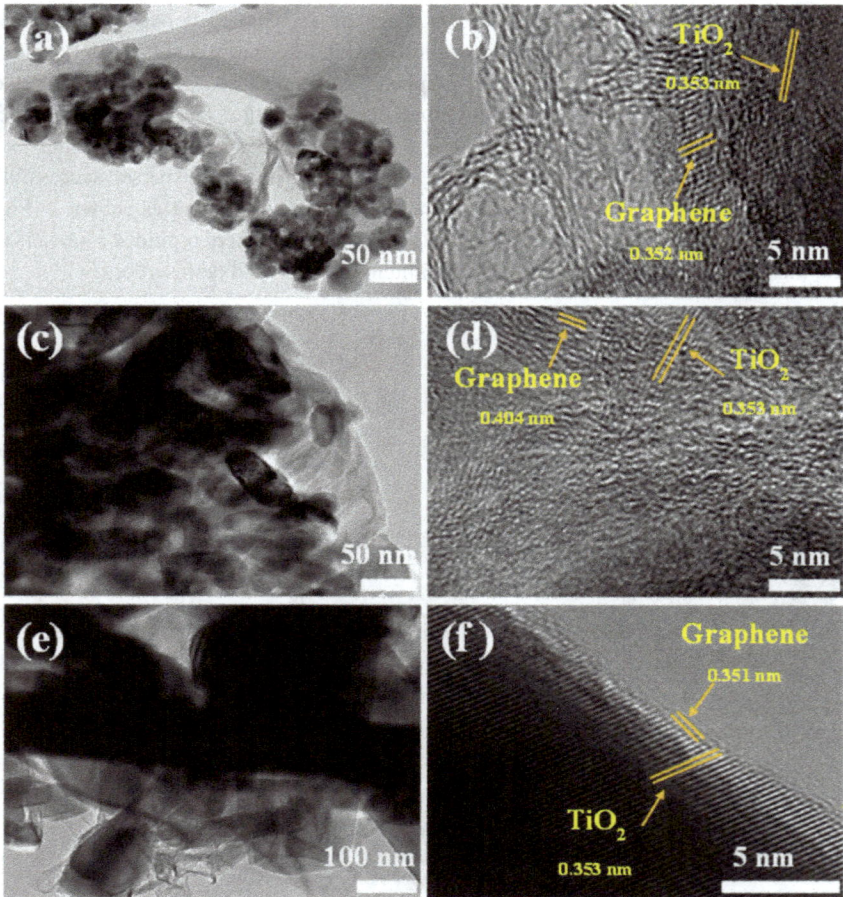

Figure 6. HR-TEM images after fabrication of nanocomposite of anatase TiO_2 with graphene (TiO_2@rGO). (a, b) TiO_2 NPAS@rGO, (c, d) TiO_2 NRDS@rGO, (e, f) TiO_2 NFBS@rGO [13].

4.1.3 XRD

TiO_2 NRDS and TiO_2 NFBS were annealed in Ar atmosphere for 1 h at 500 °C and 800 °C, respectively. XRD patterns of all TiO_2 nanostructures before fabrication of nanocomposites with graphnen are depicted in Fig. 7a. TiO_2 NRDS and TiO_2 NFBS underwent phase change into anatase crystalline form after annealing. Diffraction peaks of all nanostructures are ascribed to tetragonal anatase TiO_2 (JCPDS No. 21-1272).

After fabrication of nanocomposites TiO_2@rGO were calcinated in Ar at 400 °C for 4 h. XRD patterns of all nanocomposites after calcination are illustrated in Fig. 7b. GO sample unveiled one characteristic peak (001) interlayer spacing, reflecting oxidation of natural graphite into GO [15]. Diffraction intensities of TiO_2@rGO nanocomposites were improved after hydrothermal and calcination treatment, demonstrating increment in crystallinity of TiO_2. Notably, reflections related to graphene oxide or graphene were not observed, suggesting small percentage of carbon in these nanocomposites.

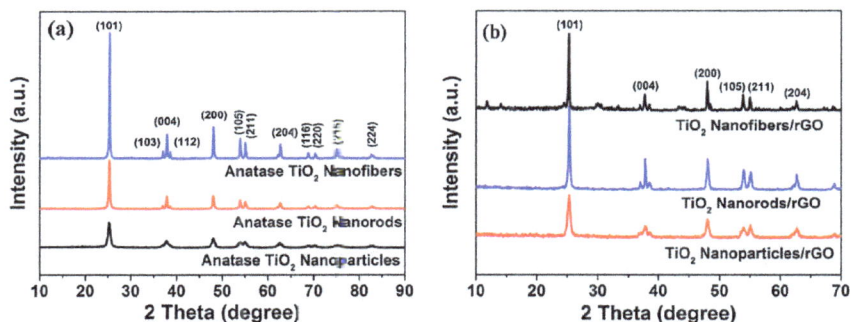

Figure 7. XRD patterns of TiO_2 nanostructures. (a) Before fabrication of nanocomposites with graphene, (b) After fabrication of nanocomposites with graphene [13].

4.1.4 Raman

To obtain further evidence about the structure of as-prepared samples, Raman spectra were collected. Results are shown in Fig. 8. Three consecutive Raman scattering peaks within 300–700 cm^{-1} corresponded to TiO_2 anatase vibrational modes [22]. The peak at around 1351 cm^{-1} corresponded to k-point phonons of A_{1g} and the peak at 1595 cm^{-1} G band originated from E_{2g} phonon of sp^2 in TiO_2@rGO nanocomposites [23]. As shown in Table 2, intensity ratios of (I_D/I_G) of all nanocomposites suggested that TiO_2 nanostructures were dispersed between graphene sheets [24]. These intensity ratios (I_D/I_G) of TiO_2@rGO nanocomposites compared to I_D/I_G of GO (0.81) indicated reduction of GO [25].

Table 2. I_D/I_G ratio of all nanocomposites characterized by Raman spectra [13].

Samples	I_D/I_G
TiO$_2$ NPAS@rGO	1.06
TiO$_2$ NRDS@rGO	1.05
TiO$_2$ NFBS@rGO	1.02

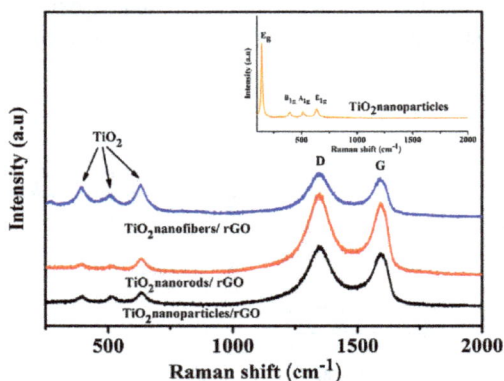

Figure 8. Raman spectra of TiO$_2$ nanostructures after fabrication of nanocomposites with graphene TiO$_2$@rGO. Inset: Raman spectra of TiO$_2$ NPAS before fabrication of nanocomposite with graphene [13].

4.1.5 BET

Table 3 indicates N$_2$ adsorption-desorption isotherms and similarly pore size distribution of all TiO$_2$ nanostructures before and after fabrication of nanocomposites with graphene. From adsorption isotherm curves, TiO$_2$ NPAS showed the highest specific surface area of 69.0 m^{2}g^{-1} along pore size distribution of 10.3 nm as illustrated in Table. 3. TiO$_2$ NFBS had the lowest surface area of 10.9 m^{2}g^{-1} with Barrett-Joyner-Halenda (BJH) pore diameter of 11.6 nm (Table 3). BET and BJH of all nanostructures of TiO$_2$ after fabricating nanocomposites with graphene are shown in Table 3. After fabricating nanocomposites with graphene, all nanostructures of TiO$_2$ improved its surface area owing to the contribution of graphene. Among TiO$_2$@rGO nanocomposites, TiO$_2$

NPAS@rGO hold the highest surface area of 97 m^2g^{-1} due to contribution of graphene as shown in Table 3. Diffusion of lithium-ions largely depends on the surface area and pore size distribution as large pore size and surface area would provide more open channels for the transport of lithium ions throughout the TiO_2@rGO nanocomposites. This is favorable for application in LIBs. Homogeneous dispersion of graphene layers among TiO_2 nanostructures can serve as a bridge which is favorable for fast insertion of lithium-ions.

Table 3. BET & BJH for all TiO_2 nanostructures and their nanocomposites with graphene [13].

Samples	BET surface area (m^2g^{-1})	BJH pore diameter (nm)	Cumulative pore volume (cm^3g^{-1})
TiO_2 NPAS	69.0	10.3	0.2113
TiO_2 NRDS	22.76	43.47	0.3010
TiO_2 NFBS	10.9	11.68	0.0215
TiO_2 NPAS@rGO	97.709	3.761	0.307
TiO_2 NRDS@rGO	69.385	3.739	0.350
TiO_2 NFBS@rGO	19.358	3.695	0.086

4.1.6 EDX

Fig. 9 shows EDX spectra of TiO_2 nanocomposites fabricated with graphene after calcination in Ar flow at 400 °C for 4 h. EDX spectra results revealed that TiO_2 nanocomposites fabricated with graphene contained elements Titanium, Carbon, and Oxygen. Particularly, there was no significant amount of Na in resulting nanocomposites, confirming purity of the sample. EDX spectra results of TiO_2 nanocomposites fabricated with graphene also indicated that actual percentages of graphene in these nanocomposites were around ~13%.

4.2 Electrochemical Testing

Electrochemical performances of as-prepared samples were tested as negative materials for LIBs using lithium metal as reference electrodes fabricated into CR2032 coin cells. Electrochemical performances of TiO_2@rGO nanocomposites were investigated within potential between of 1.0 and 3.0 V vs Li/Li$^+$. Results of specific capacity and rate capability relays on many aspects, including size and morphology of TiO_2, type of bonding among TiO_2 and graphene, and surficial properties (such as surface area, pore size, porosity) of TiO_2 nanostructures. Fig. 10a-10c display initial four cycles of charge/discharge plateaus of TiO_2@rGO nanocomposites at a C-rate of 0.5 C. In Figs. 10a and 10b, plateaus at about ~1.7 V and 2.0 V vs Li/Li^+ were horizontal typical lithium insertion and viceversa. They could be ascribed to lithium insertion and extraction between TiO_2 and $Li_{0.5}TiO_2$ [26–28]. The small capacity decay in initial few cycles owing to irreversible reactions and formation of byproducts [29]. However, the profile for TiO_2 NFBS@rGO in Fig. 10(c) for lithium ion storage was distinct. The slope profile was likely due to uniform results with a small capacity fade. Additionally, TiO_2 NFBS@rGO with small surface area and pore volume prompted a capacitive behavior owing to improved lithium ion storage due to lithiation mechanism.

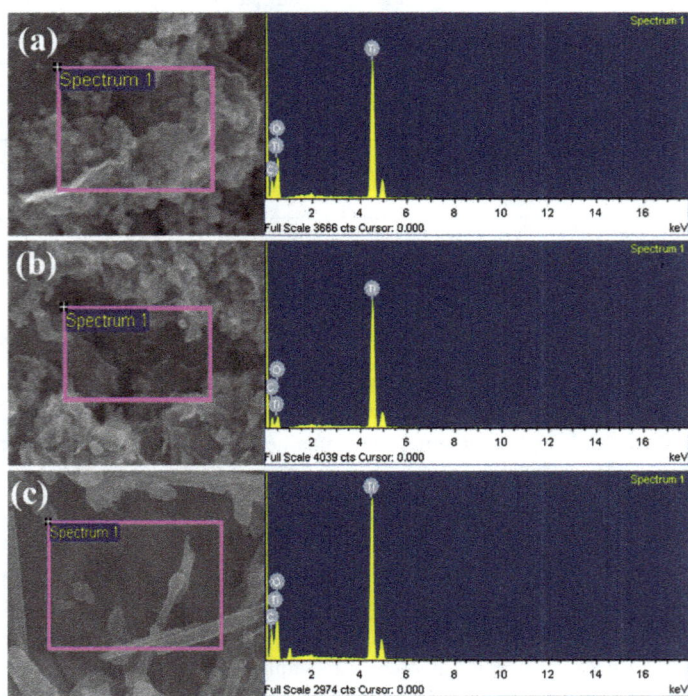

Figure 9. EDX spectra of TiO_2 nanocomposites with graphene TiO_2@rGO after calcination in Ar at 400 °C for 4 h. (a) TiO_2 NPAS@rGO, (b) TiO_2 NRDS@rGO, (c) TiO_2 NFBS@rGO [13].

Rate capabilities at various C rates for nanocomposites of TiO_2@rGO are shown in Fig. 10d. Rate capabilities of TiO_2 NRDS@rGO at 10 C and 20 C were higher than those of TiO_2 NPAS@rGO because TiO_2 NAPS@rGO underwent high resistive path at high C-rate. However, TiO_2 NRDS/rGO supplied a low resistive path compared to TiO_2 NPAS@rGO because rate capabilities of TiO_2 NRDS@rGO were higher at higher C-rate. TiO_2 NRDS@rGO and TiO_2 NFBS@rGO had rate capabilities of 42 mAhg^{-1} and 24 mAhg^{-1}, respectively, at 20 C. However, after applying a high C-rate of 20 C, TiO_2 NFBS@rGO delivered a rate capability of 64 mAhg^{-1} at C-rate of 0.5 C. This might be due to a high contact area that could lead to large electrode/electrolyte contact area. The high rate capacity TiO_2 NRDS@rGO, could be as a result of lower aggregation and well mixing of TiO_2 NRDS among graphene that might retain a high contact area to supply more reactive channels as a result of specific C-rate. Interestingly, TiO_2 NPAS@rGO after application of a high C-rate, rate capability at C-rate of 0.5 C recovered to 109 mAhg^{-1}, unveiling a low structural stability owing to phase change from anatase into rutile [28,30].

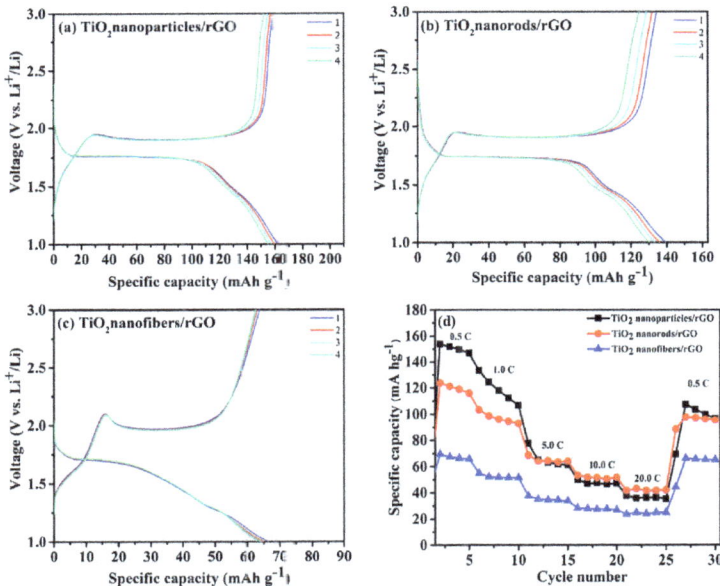

Figure 10. Charge-discharge profiles at 0.5 C (1C = 168 mAhg^{-1}). (a) TiO_2 NPAS@rGO, (b) TiO_2 NRDS@rGO, (c) TiO_2 NFBS@rGO, (d) Rate performance at different current densities of TiO_2 NPAS@rGO, TiO_2 NRDS@rGO, and TiO_2 NFBS@rGO [13].

Cyclic performances of all TiO_2@rGO nanocomposites for 100 cycles at C-rate of 10 C were estimated. Results are shown in Fig. 11. In the first 20 cycles, TiO_2 NPAS@rGO underwent high capacity loss and exhibited a discharge capacity of 52 mAhg^{-1} for 100 cycles. Solid-electrolyte interface (SEI) layer formation could be a main reason of capacity fade in initial cycles. TiO_2 NRDS@rGO delivered a discharge capacity of 53 mAhg^{-1} at C-rate of 10 C for 100 cycles, demonstrating good retention capability. TiO_2 NRDS offered low resistive path at high C-rate, demonstrating that TiO_2 NRDS would be quite reversible for lithiation-delithiation. TiO_2 NFBS@rGO displayed an initial discharge capacity of 26 mAhg^{-1}. After 100 cycles, they delivered a capacity of 30 mAhg^{-1}. Gradual increase in capacity of electrode over time was caused by the mechanism of Li-insertion and extraction process which would gently increase its porosity, leading to more open channels and more lithium storage.

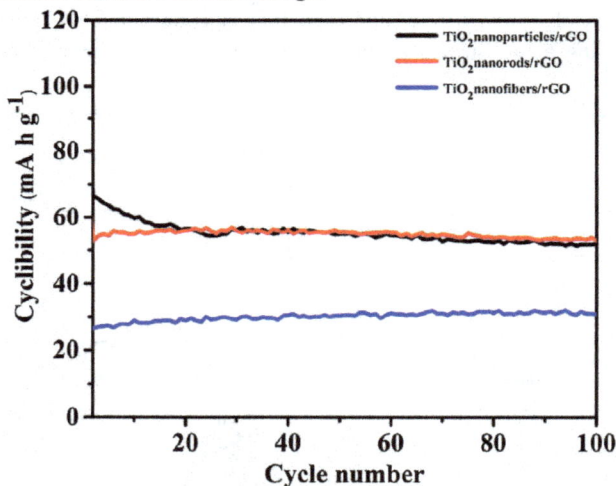

Figure 11. Discharge capacities of all TiO₂@rGO nanocomposites for 100 cycles at 10 C [13].

5.   Characterization of coaxial SnO_2@TiO_2 nanotube hybrids

5.1   Coaxial SnO_2@TiO_2 nanotube hybrids

To improve quantum efficiency, conductivity and discharged capacity of TiO_2 nanotube arrays (TNTs) were encapsulated uniformly through electrochemical fabrication with

Materials Research Forum LLC
https://doi.org/10.21741/9781644900918-1

SnO_2 including exterior-interior space. To investigate Li^+ storage capacity, synthesized hybrid material was applied as negative material for LIBs.

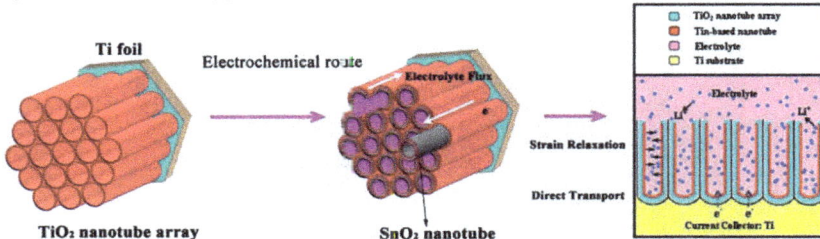

Figure 12. Schematic illustration of SnO_2@TiO_2 coaxial nanotube hybrids (SnO_2@TNTs) and its benefit as an electrode material for Li-ion storage [31].

5.1.1 SEM & TEM

Morphological analyses results of coaxial Sn@TiO_2 nanotubes (Sn@TNTs) and SnO_2@TiO_2 nanotubes (SnO_2@TNTs) synthesized via electrochemical fabrication method are illustrated in Figs. 13a-13e. These synthesized materials showed a hierarchical structure, in which TNTs were encapsulated uniformly from inside-outside in a similar fashion. It was noticed that some walls of SnO_2 were not firmly attached with TNTs. In spite of that, a definite double-shell core structure was evident could be seen in Fig. 13(a). The inset of Fig. 13(a) confirms successful annealing at 500 °C for 90 min in air atmosphere. After annealing, desired crystalline phase was achieved without sacrificing the original morphology.

From TEM images shown in Fig. 13(b), the thickness of doubled-shelled core structure can be evaluated. Fig. 13(b) revealed that the thickness of multilayer was around 40 nm and average thicknesses of TNTs and SnO_2 were in the range of 10 to 20 nm. Apparently, there was an inner space between walls of TNTs and metallic Sn possibly caused by hydrogen evolution ($2H^+ + 2e^- \rightarrow H_2$) during the deposition of Sn at high negative potential. Irrespective of reaction time, thickness and attachment of SnO_2 layer depended on the concentration of Sn^{+2} as shown in Fig. 13c. TEM images further confirmed the uniform tin deposition on TNTs. As shown in insets of Figs. 13b and 13d, layer distribution and thickness of SnO_2 were uniform; revealing that Sn deposition in the inner space of TNTs was initiated simultaneously. Besides, boundary walls between TNTs and tin were not well distinguishable as shown in Fig. 13d. This might be because good fusion between the interface of TNTs and Sn could be achieved after particular heat

Lithium-ion Batteries - Materials and Applications
Materials Research Foundations **80** (2020) 1-27

Materials Research Forum LLC
https://doi.org/10.21741/9781644900918-1

treatment. The spot and ring patterns in selected area electron diffraction (SAED) image indicated the existence of both materials (anatase and cassiterite). In Fig. 13e, the lattice fringes with a period of (d = 0.33 nm) corresponds to the (110) lattice planes of SnO_2.

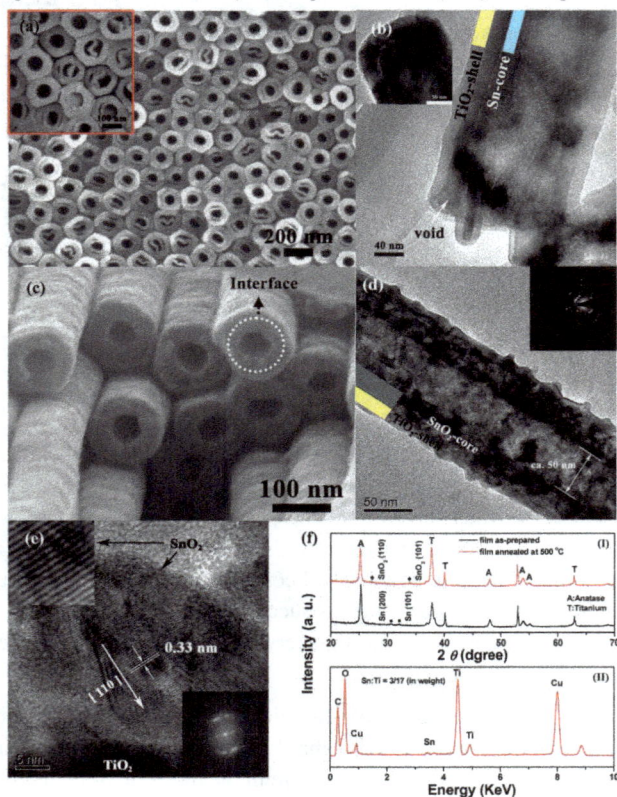

Figure 13. Characterization of coaxial bimodal SnO_2@TNTs and Sn@TNTs synthesized by sequential electrochemical fabrication. (a) Top look of Sn nanotubes encapsulated in TNTs and after annealing the sample (inset), (b) TEM images of coaxial Sn@TNTs bottom view, mouth view (inset), (c) SnO_2@TNTs images with tightly encapsulated SnO_2 monolayer, (d) Segmented TEM image of SnO_2@TNTs, with inset SAED pattern indicating that the sample is polycrystalline, (e) Upper region HR-TEM image of SnO_2@TNTs and Fast Fourier Transform (FFT) revealing the existence of single crystalline SnO_2. (f) Before and after annealing of similar coaxial nanotube XRD patterns (I) and EDX patterns observed from SnO_2@TNTs (II) [31].

5.1.2 XRD

Fig. 13f(I) shows XRD patterns of samples before and after annealing at 500 °C for 90 min in air atmosphere. After deposition of Sn on TNTs, most of diffraction peaks could be well indexed to the anatase phase of TiO_2 (JCPDS no. 21-1272). However, the two diffraction peaks at 30.6° and 32° corresponded to β-Sn. After annealing at 500 °C for 90 min in air atmosphere, the existence of two low diffraction peaks at 26.8° and 34.05° was evident. Most of metallic Sn was transformed into cassiterite SnO_2 (JCPDS no. 41-1445).

5.1.3 EDX

EDX spectra of individual SnO_2@TNTs are shown in Fig. 13f(II). Peaks in the range of 3.4 - 4.1 KeV indicated the existence of Sn element. The weight ratio of Sn : Ti was 3 : 17.

5.1.4 Electrochemical testing

Fig. 14 shows the 5th cycle charge/discharge curves of pristine TNTs, Sn@TNTs, and SnO_2@TNTs. The profile of SnO_2@TNTs demonstrated low voltage difference among charge and discharge profile compared to that of pristine TNTs owing to its low energy barrier for the conversion reaction. Notably prominent plateaus at around 1.75 V and 2.0 V were assigned to the biphasic equilibrium of TiO_2, reflecting the fusion of SnO_2 nanotubes. The fusion of SnO_2 nanotubes also participates to harvest the TiO_2 capacity and offers Li-ions to move through the SnO_2 layer without much hindrance at the stated current density.

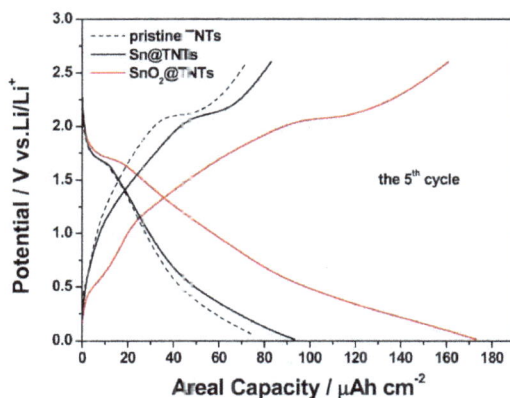

Figure 14. Fifth cycle galvanostatic charge/discharge profile of pristine TNTs, Sn@TNTs, and SnO_2@TNTs [31].

Lithium-ion Batteries - Materials and Applications   Materials Research Forum LLC
Materials Research Foundations 80 (2020) 1-27   https://doi.org/10.21741/9781644900918-1

Fig. 15 depicts discharge capacity for 50 cycles of pristine TNTs, Sn@TNTs, and SnO$_2$@TNTs at current density of 100 μA cm^{-2} (film thickness: ca.1.8 μm). SnO$_2$@TNTs electeode exhibited the first step discharge capacity of 469.8 μA h cm^{-2}. The irreversible fade in capacity in the first few cycles owing to the formation of SEI layer as a result of irreversible reaction between Li-ion and electrolyte solvent molecules. Nevertheless, in subsequent cycles, coulombic efficiency of more than 94 % was obtained, showing reversible capacity of ~113 μA h cm^{-2} after 50 cycles which was higher than that of pristine TNTs (51.6 μA h cm^{-2}). Furthermore, unlike Sn@TNTs, the significant increase in capacity of annealed sample (SnO$_2$@TNTs) also attributed to enhanced fusion between two components. Additionally, the fade in capacity of composite electrode indicated that pulverization of Sn material in the nano-hetrostructure form was greatly increased.

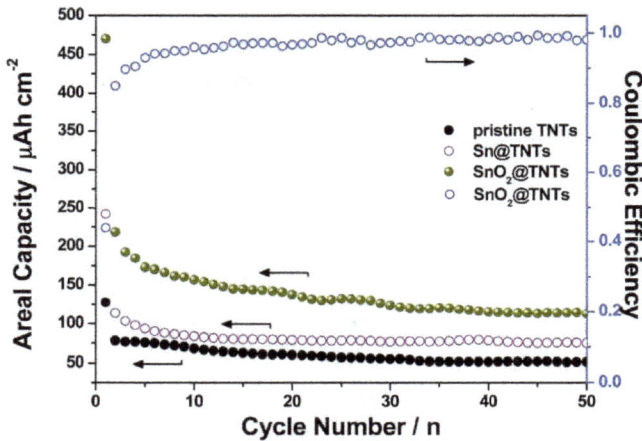

Figure 15. Cyclic performance and coulombic efficiency vs. cycle number/n for pristine TNTs, Sn@TNTs, and SnO$_2$@TNTs at current density of 100 μA cm^{-2} [31].

It is notable to consider the interfacial chemical reaction at high temperature between oxygen from the titanium dioxide lattice and metallic Sn. In this case, Ti undergoes partial reduction from Ti^{4+} to Ti^{+3} or Ti^{2+}, which promotes an oxygen deficient environment among TiO$_2$ and metallic Sn interface layer. The electronic conductivity of TNTs increases with the increase in defect sites, which results in favorable charge transport. To further verify the improvement in conductivity of SnO$_2$@TNTs impedance measurements were performed, as depicted in Fig. 16. In Fig. 16, it is evident that the charge-transfer resistance (R_{ct}) of SnO_2@TNTs in a wide potential window is smaller than pristine TiO_2. Furthermore, during the discharge process the decline in R_{ct} of SnO_2@TNTs should be attributed to the slow reduction of Ti^{4+} to Ti^{+3} or Ti^{2+}, accompanied by the lithiation among TNTs. Which in turn, reduces the mobility gap and electrical resistance.

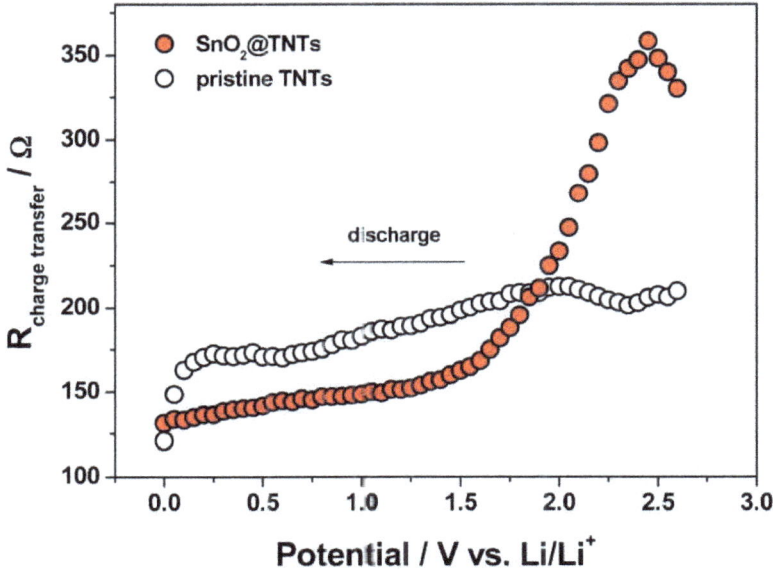

Figure 16. Electrochemical impedance data of coaxial SnO_2@TNTs and pristine TNTs [31].

Conclusion

In this chapter, nanocomposites of TiO_2 (nanoparticles, nanorods, nanofibers) with graphene (TiO_2@rGO) were synthesized using hydrothermal and calcination treatment. Coaxial SnO_2@TNTs were successfully assembled through electrochemical fabrication technique using pristine TNTs and metallic Sn. Physicochemical characteristics of as-synthesized products TiO_2@rGO and SnO_2@TNTs were analyzed using SEM, HR-TEM, XRD, RAMAN, BET, and EDX spectroscopy. Their electrochemical performances were carried out as negative materials for LIBs by fabricating half-cell for as-synthesized products using lithium metal as a counter electrode. Based on our results, the following conclusions were drawn.

I.   As a consequence of hydrothermal and calcination treatment, well mixing and uniform dispersion of TiO_2 nanostructure between graphene layers were obtained.

II. TiO_2 NPAS@rGO delivered a rate capability of 155 mAhg^{-1} at C-rate of 0.5 C. After applying a high C-rate of 20 C, rate capability was 109 mAhg^{-1} at 0.5 C, showing capacity loss of 30 %.

III. TiO_2 NRDS@rGO exhibited cyclic reversibility of 53 mAhg^{-1} after 100 cycles at 10 C with almost ~100 reversibility, indicating strong bonding between TiO_2 NRDS and graphene.

IV. TiO_2 NFBS@rGO unveiled cycling reversibility of 30 mAhg^{-1} after 100 cycles at 10 C with a remarkable retention capacity owing to its expandable porosity during lithiation and delithiation.

V.   Coaxial SnO_2@TNTS monohybrid were fabricated successfully using electrochemical fabrication methodology, resulting in uniform deposition of metallic-Sn inside/outside of TNTs.

VI. Enhanced fusion was achieved between TNTs and metallic Sn by annealing at 500 °C for 90 min in air atmosphere.

VII. SnO_2@TNTs displayed reversible capacity of 113 µA h cm^{-2} after 50 cycles at current density of 100 µA cm^{-2} compared to pristine TNTs (51.6 µA h cm^{-2}). The enhanced capacity was attributed to the addition of high capacity component and subsequent sample annealing.

Acknowledgement

We acknowledge financial support from BISA Research Grant of Keimyung University in 2019.

References

[1] K. Wang, X. Li, J. Chen, Surface and interface engineering of electrode materials for lithium-ion batteries, Adv. Mater. 27 (2015) 527-545. https://doi.org/10.1002/adma.201402962

[2]   P. Zheng, T. Liu, Y. Su, L. Zhang, S. Guo, TiO_2 nanotubes wrapped with reduced graphene oxide as a high-performance anode material for lithium-ion batteries, Sci. Rep. 6 (2016) 36580. https://doi.org/10.1038/srep36580

[3] M. Minella, D. Versaci, S. Casino, F. Di Lupo, C. Minero, A. Battiato, N. Penazzi, S. Bodoardo, Anodic materials for lithium-ion batteries: TiO_2-rGO composites for high power applications, Electrochim. Acta 230 (2017) 132–140. https://doi.org/10.1016/j.electacta.2017.01.190

[4] P. Xiong, L. Peng, D. Chen, Y. Zhao, X. Wang, G. Yu, Two-dimensional nanosheets based Li-ion full batteries with high rate capability and flexibility, Nano Energy 12 (2015) 816-823. https://doi.org/10.1016/j.nanoen.2015.01.044

[5] D. Ma, Z. Cao, A. Hu, Si-based anode materials for Li-ion batteries: A mini review, Nano-Micro Letters 6 (2014) 347-358. https://doi.org/10.1007/s40820-014-0008-2

[6] S. Goriparti, E. Miele, F. De Angelis, E. Di Fabrizio, R.P. Zaccaria, C. Capiglia, Review on recent progress of nanostructured anode materials for Li-ion batteries, J. Power Sources 257 (2014) 421-443. https://doi.org/10.1016/j.jpowsour.2013.11.103

[7] L. Bai, F. Fang, Y. Zhao, Y. Liu, J. Li, G. Huang, H. Sun, A sandwich structure of mesoporous anatase TiO_2 sheets and reduced graphene oxide and its application as lithium-ion battery electrodes, RSC Adv. 4 (2014) 43039-43046. https://doi.org/10.1039/C4RA04979A

[8] N. Nitta, F. Wu, J.T. Lee, G. Yushin, Li-ion battery materials: Present and future, Mater. Today 18 (2015) 252-264. https://doi.org/10.1016/j.mattod.2014.10.040

[9] X. Zheng, J. Li, A review of research on hematite as anode material for lithium-ion batteries, Ionics 20 (2014) 1651-1663. https://doi.org/10.1007/s11581-014-1262-5

[10] V.J. Babu, S. Vempati, T. Uyar, S. Ramakrishna, Review of one-dimensional and two-dimensional nanostructured materials for hydrogen generation, Phys. Chem. Chem. Phys. 17 (2015) 2960-2986. https://doi.org/10.1039/C4CP04245J

[11] X. Li, Y. Zhang, T. Li, Q. Zhong, H. Li, J. Huang, Graphene nanoscrolls encapsulated TiO_2 (B) nanowires for lithium storage, J. Power Sources 268 (2014) 372-378. https://doi.org/10.1016/j.jpowsour.2014.06.056

[12] J. Jin, S. Huang, J. Liu, Y. Li, D. Chen, H. Wang, Y. Yu, L. Chen, B. Su, Design of new anode materials based on hierarchical, three dimensional ordered macro-mesoporous TiO_2 for high performance lithium ion batteries, J. Mater. Chem. A 2 (2014) 9699-9708. https://doi.org/10.1039/c4ta01775g

[13] H. Jamal, B. Kang, H. Lee, J. Yu, C. Lee, Comparative studies of electrochemical performance and characterization of TiO_2/graphene nanocomposites as anode

materials for Li-secondary batteries, Journal of Industrial and Engineering Chemistry 64 (2018) 151-166. https://doi.org/10.1016/j.jiec.2018.03.012

[14]   L. Yu, Z. Wang, L. Zhang, H.B. Wu, X.W.D. Lou, TiO_2 nanotube arrays grafted with Fe_2O_3 hollow nanorods as integrated electrodes for lithium-ion batteries, Journal of Materials Chemistry A 1 (2013) 122-127. https://doi.org/10.1039/C2TA00223J

[15]   Z. Xiu, X. Hao, Y. Wu, Q. Lu, S. Liu, Graphene-bonded and-encapsulated mesoporous TiO_2 microspheres as a high-performance anode material for lithium ion batteries, J. Power Sources 287 (2015) 334-340. https://doi.org/10.1016/j.jpowsour.2015.04.086

[16]   G. Zhu, Y. Wang, Y. Xia, Ti-based compounds as anode materials for Li-ion batteries, Energy & Environmental Science 5 (2012) 6652-6667. https://doi.org/10.1039/c2ee03410g

[17]   W. Wei, G. Oltean, C. Tai, K. Edström, F. Björefors, L. Nyholm, High energy and power density TiO_2 nanotube electrodes for 3D Li-ion microbatteries, Journal of Materials Chemistry A 1 (2013) 8160-8169. https://doi.org/10.1039/c3ta11273j

[18]   Y. Liu, Y. Yang, Recent progress of TiO_2-based anodes for Li ion batteries, Journal of Nanomaterials 2016 (2016) 2. https://doi.org/10.1155/2016/8123652

[19]   P. Bottke, Y. Ren, I. Hanzu, P.G. Bruce, M. Wilkening, Li ion dynamics in TiO_2 anode materials with an ordered hierarchical pore structure–insights from ex situ NMR, Physical Chemistry Chemical Physics 16 (2014) 1894-1901. https://doi.org/10.1039/C3CP54586E

[20]   D. Yuan, W. Yang, J. Ni, L. Gao, Sandwich structured $MoO_2@TiO_2@CNT$ nanocomposites with high-rate performance for lithium ion batteries, Electrochim. Acta 163 (2015) 57-63. https://doi.org/10.1016/j.electacta.2015.02.149

[21]   C. Yang, Z. Wang, T. Lin, H. Yin, X. Lü, D. Wan, T. Xu, C. Zheng, J. Lin, F. Huang, Core-shell nanostructured "black" rutile titania as excellent catalyst for hydrogen production enhanced by sulfur doping, J. Am. Chem. Soc. 135 (2013) 17831-17838. https://doi.org/10.1021/ja4076748

[22]   T. Ohsaka, F. Izumi, Y. Fujiki, Raman spectrum of anatase, TiO_2, J. Raman Spectrosc. 7 (1978) 321–324. https://doi.org/10.1002/jrs.1250070606

[23]   H. Liu, W. Li, D. Shen, D. Zhao, G. Wang, Graphitic carbon conformal coating of mesoporous TiO_2 hollow spheres for high-performance lithium ion battery anodes, J. Am. Chem. Soc. 137 (2015) 13161-13166. https://doi.org/10.1021/jacs.5b08743

[24]   X. Tong, M. Zeng, J. Li, F. Li, UV-assisted synthesis of surface modified mesoporous TiO_2/G microspheres and its electrochemical performances in lithium ion batteries, Appl. Surf. Sci. 392 (2017) 897–903. https://doi.org/10.1016/j.apsusc.2016.09.094

[25]   A. Razzaq, C.A. Grimes, S. Il In, Facile fabrication of a noble metal-free photocatalyst: TiO_2 nanotube arrays covered with reduced graphene oxide, Carbon N. Y. 98 (2016) 537–544. https://doi.org/10.1016/j.carbon.2015.11.053

[26]   S. Yang, X. Feng, K. Müllen, Sandwich-like, graphene-based titania nanosheets with high surface area for fast lithium storage, Adv. Mater. 23 (2011) 3575–3579. https://doi.org/10.1002/adma.201101599

[27]   H. Liu, K. Cao, X. Xu, L. Jiao, Y. Wang, H. Yuan, Ultrasmall TiO_2 Nanoparticles in Situ Growth on Graphene Hybrid as Superior Anode Material for Sodium/Lithium Ion Batteries, ACS Appl. Mater. Interfaces. 7 (2015) 11239–11245. https://doi.org/10.1021/acsami.5b02724

[28]   N.D. Petkovich, B.E. Wilson, S.G. Rudisill, A. Stein, Titania-carbon nanocomposite anodes for lithium ion batteries - Effects of confined growth and phase synergism, ACS Appl. Mater. Interfaces. 6 (2014) 18215–18227. https://doi.org/10.1021/am505210c

[29]   Y. Xie, J. Song, P. Zhou, Y. Ling, Y. Wu, Controllable Synthesis of TiO_2/Graphene Nanocomposites for Long Lifetime Lithium Storage: Nanoparticles vs. Nanolayers, Electrochim. Acta. 210 (2016) 358–366. https://doi.org/10.1016/j.electacta.2016.05.157

[30]   R. Mo, Z. Lei, K. Sun, D. Rooney, Facile synthesis of anatase TiO_2 quantum-dot/graphene-nanosheet composites with enhanced electrochemical performance for lithium-ion batteries, Adv. Mater. 26 (2014) 2084–2088. https://doi.org/10.1002/adma.201304338

[31]   X. Wu, S. Zhang, L. Wang, Z. Du, H. Fang, Y. Ling, Z. Huang, Coaxial SnO_2@TiO_2 nanotube hybrids: From robust assembly strategies to potential application in Li^+ storage, J. Mater. Chem. 22 (2012) 11151–11158. https://doi.org/10.1039/c2jm30885a

Lithium-ion Batteries - Materials and Applications   Materials Research Forum LLC
Materials Research Foundations **80** (2020) 28-62   https://doi.org/10.21741/9781644900918-2

Chapter 2

A Brief History of Conducting Polymers Applied in Lithium-ion Batteries

Suzhe Liang[1,2], Yonggao Xia[2,3*], Peter Müller-Buschbaum[1,4*], Ya-Jun Cheng[2,5*]

[1]Lehrstuhl für Funktionelle Materialien, Physik-Department, Technische Universität München, James-Franck-Str. 1, 85748, Garching, Germany

[2]Ningbo Institute of Materials Technology & Engineering, Chinese Academy of Sciences, 1219 Zhongguan West Rd, Zhenhai District, Ningbo, Zhejiang Province, 315201, PR China

[3]Center of Materials Science and Optoelectronics Engineering, University of Chinese Academy of Sciences, 19A Yuquan Rd, Shijingshan District, Beijing 100049, P. R. China

[4]Heinz Maier-Leibnitz Zentrum (MLZ), Technische Universität München, Lichtenbergstr. 1, 85748, Garching, Germany

[5]Department of Materials, University of Oxford, Parks Rd, OX1 3PH, Oxford, United Kingdom

*xiayg@nimte.ac.cn, muellerb@ph.tum.de, chengyj@nimte.ac.cn

Abstract

As the most important portable power source at present, lithium-ion batteries still suffer from several problems due to lack of perfect electrode materials. Conducting polymers, owning both good electrical and physical properties, not only can composite with conventional electrode materials to improve their electrochemical performance but also are able to separately used as electrodes. In this chapter, the related studies of conducting polymer applied in electrodes will be reviewed by a chronological approach. With the unique perspective, this chapter will shed light on the future trends of the studies on applications of conducting polymers in lithium-ion batteries.

Keywords

Conducting Polymer, Brief History, Cathode, Anode, Lithium-Ion Battery

Contents

1.   Introduction

Since the successful commercialization in the 1990s, the lithium-ion battery (LIB) has become one of the most important power storage media in contemporary era [1-4]. Because of high output voltage, high energy density, long cycle life, and good rate capability, lithium-ion batteries have dominated the portable electronics market and are expanding the electric vehicles market in an astonishing speed [5, 6]. However, the first-generation lithium-ion batteries no longer meet the gradually increasing demands of higher energy density and more diverse operation environments [6, 7]. Generally, a lithium-ion battery consists of the copper current collector, anode, electrolyte, separator, cathode, aluminum current collector, and packing materials. Among these, the kind and form of electrode materials can directly determine the final electrochemical performance and applicability of batteries. At present, lithium transition metal oxides or phosphates, such as $LiCO_2$, $LiMnO_4$, $LiFePO_4$, or $Li(NiCoMn)O_2$, constitute the cathode materials of commercialized LIBs, while graphite-based materials act as anodes [8]. The relative-low theoretical capacities of these existing electrode materials limit the gravimetric energy density of LIBs under 250 Wh kg^{-1}. To pursue higher energy density, it is very urgent to implement an update revolution for the global LIB market. Naturally, designing and fabricating high-performance electrode materials is the key point of this revolution. For the cathode materials, how to improve the capacity and Coulombic efficiency is the hotpot in both academia and industry, because many kinds of cathode materials have not yet fully realized their potentials [9-11]. In terms of anode materials, various high-capacity materials have been developed in academia, such as silicon (Si), phosphorus (P), germanium (Ge), tin (Sn) and antimony (Sb)-based materials [12-14]. However, these alloy-type anode materials suffer from sever volumetric changes during charge/discharge processes, which leads to fast electrode pulverization and capacity fading. This is the major obstacle against the practical applications of these anode materials. In addition, the emergence of flexible and wearable electronic devices also needs the new-type power sources [15-18]. In recent years, organic/inorganic hybrid electrode materials attract much attention due to their unique advantages compared to the conventional inorganic electrodes [19-21]. The existence of an organic component can provide several new characteristics to the traditional electrodes. Firstly, some organic molecules have a higher theoretical capacity than that of traditional inorganic cathode materials, which benefits to achieve higher energy density [22]. Secondly, the flexible and plastic organic materials not only act as a buffer matrix to alleviate the volume expansion of alloy-type anodes but also endow the flexibility to the electrodes [23]. Moreover, the utilization of organic materials can relieve the pressure of large-scale use of transition metal as electrode materials, leading to more sustainable and environment-friend development [24]. Among numerous organic materials, conducting polymers (CPs) are ideal materials for hybrid electrodes of lithium-ion batteries, due to good electrical conductivity and basic properties of organic materials [20].

Conducting polymers or intrinsically conductive polymers, as the name implies, are a kind of polymers which present electric conductivity [25, 26]. Their conductive nature is derived from altering single and double carbon-carbon bonds in polymer chains. Conducting polymer belongs to conjugated polymer, in which the sp^2p_z configuration of carbon orbitals (in π-bonding) and the overlap of successive carbon atoms' orbitals along the backbone, leading to electron delocalization along the backbone of the polymer [27]. The delocalized electrons can freely move along the backbone of the chain, resulting in the conductivity of the polymer. In 1976, Alan MacDiarmid, Hideki Shirakawa and Alan J. Heeger discovered conducting polymers and introduced electrochemical doping into conducting polymers to control the conductivity which ranged from insulative to fully conductive in 1980 [27-29]. Therefore, they were awarded the Noble Prize in Chemistry in 2002 "for the discovery and development of conductive polymers". After about 40-years development, conducting polymers have been successfully applied in various fields, including transistors, sensors, memories, actuators/artificial muscles, supercapacitors, and lithium-ion batteries [30-35].

Combined flexibility of polymer and conductivity of the metal, conducting polymers have played a significant role in constructing advanced LIB electrode materials and gained great progress. On the one hand, CPs can be directly applied as electrodes of Li-ion batteries, due to the conversion redox mechanism [36]. On the other hand, more commonly, CPs are composited with conventional inorganic materials to form hybrid electrodes to enhance electrochemical performance. A verity of conductive polymers have been developed for the applications of electrodes, including polyacetylene (PA), polyaniline (PANI), polypyrrole (PPy), polythiophene (PTh), poly(para-phenylene) (PPP), poly(para-phenylene vinylene) (PPV), polyfuran (PF), and poly(3,4-ethlenedioxythiophene) (PEDOT) [19, 36, 37]. The molecular structures of these typical conducting polymers are displayed in Fig. 1. In this chapter, the progress of the studies about applications of conducting polymers in lithium-ion batteries will be reviewed in a chronological style. There will be two major sections which try to draw outlines of the development history of CPs in LIB cathodes and anodes, respectively. In each section, the related studies will be divided into several timeframes, where the different developing trends of CPs will be discussed and summarized.

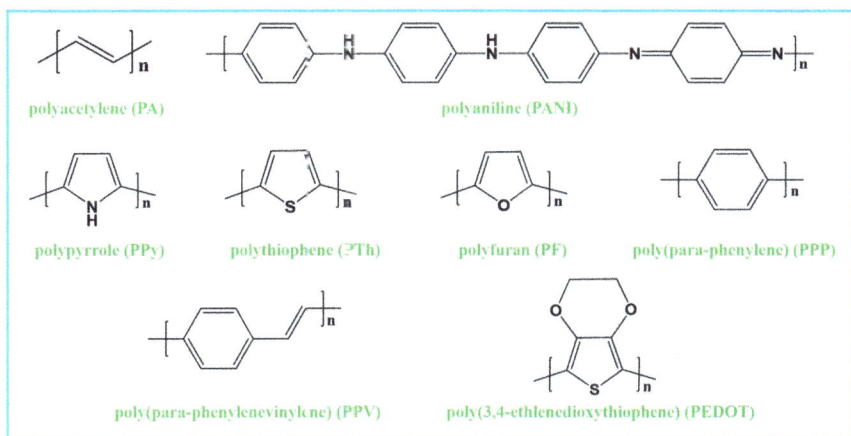

Figure 1   Molecular structures of typical conducting polymers.

2.   Applications on cathode materials

2.1   Before 2000: Emergence stage

After the discovery of conducting polymers in 1976, Alan J. Heeger *et al.* then applied them in energy storage/conversion fields [38]. At that time, the lithium-ion batteries were still not commercialized, thus most of studies on CPs focused on their applications in rechargeable Li metal batteries [38–48]. In 1981, Heeger and MacDiarmid found that the conductivity of polyacetylene (PA) could be controlled from semiconducting to the metallic regime through an electrochemical doping process [38]. Using a solution of $LiClO_4$ in propylene carbonate and a lithium cathode, PA was electrochemically doped to form $[CH^{+y}(ClO4)_y^-]$ (y=0-0.06) which had conductivities up to *ca.* 10^3 Ω^{-1} cm^{-1}.

According to this result, the $[CH^{+y}(ClO4)_y^{-}]$ film could be used as cathode-active material in lightweight rechargeable storage batteries. Based on the following discharge reaction (Eq.1),

$$[CH^{+0.06}(ClO4)_{0.06}^{-}]_x + 0.06x\text{Li} \rightarrow (CH)_x + 0.06x\text{LiCiO}_4 \qquad (1)$$

A 6 % doped $[CH^{+y}(ClO4)_y^{-}]$ film exhibited an open-circuit voltage of 3.7 V and an initial short-circuit current of 25 mA. The as-prepared $(CH)_x/\text{LiCiO}_4/\text{Li}$ rechargeable storage battery could reach an energy density of about 176 Wh kg^{-1}. To our best knowledge, this is the first work which applied conducting polymers in energy storage devices. This work demonstrated the feasibility of using CPs to fabricate lightweight rechargeable batteries and the potential possibility of various applications of CPs.

In following years, other conducting polymers were discovered and applied as cathode materials in Li-based batteries, including polythiophene (PTh) [39], polypyrrole (PPy) [40, 41, 45-47], and polyaniline (PANI) [42, 45]. In 1998, Rhee *et al.* used PPy film as cathode and UV curved polymer (poly(ethylene glycol) diacrylate, PEGDA) as electrolyte to fabricate a flexible Li-based battery [47]. The Li/PEGDA/PPy battery exhibited the upper cut-off and lower cut-off voltages of 4.0 V and 2.8 V, respectively. Besides, this battery could reach the initial discharge and charge capacities of 38 mAh cm^{-2} and 51 mAh cm^{-2}, respectively, at a current density of 0.2 mA cm^{-2}.

2.2   2000-2006: Preliminary stage

When stepped into the new century, the rising of lithium-ion batteries provided more opportunities for conducting polymers to extend their application boundaries. In this period, conducting polymer-based hybrid materials appeared to be applied as the cathode of Li-ion or Li-metal batteries [49-55]. Meanwhile, there was still a part of studies focused on the synthesis of CPs-based cathodes for Li-based batteries and their electrochemical reaction mechanisms [56-61]. Up to 2001, spinel LiMn$_2$O$_4$ has been regarded as one of the most promising cathode candidates of LIBs and also attracted much attention [62-64]. However, this material still had some problems to be solved, such as its poor cycling stability at room temperature [62]. In 2002, Mastragostino's team prepared Li$_{1.03}$Mn$_{1.97}$O$_4$ spinel by sol-gel method and then coated poly(3,4-ethylenedioxythiophene) on the particles to obtain Li$_{1.03}$Mn$_{1.97}$O$_4$/PEDOT composite cathode material [49]. The conductive nature of PEDOT could improve the electronic conductivity of the electrode. In addition, the electroactive potential range of PEDOT was same as that of Li$_{1.03}$Mn$_{1.97}$O$_4$, and the plastic property of PEDOT allowed it to act as binder in the electrode. Compared to the conventional $Li_{1.03}Mn_{1.97}O_4/C$ composite cathode, the $Li_{1.03}Mn_{1.97}O_4/PEDOT$ electrode exhibited almost the same initial capacities at different current densities. In the following studies of Mastragostino's team, the electrochemical performance of $Li_xMn_{3-x}O_4/PEDOT$ hybrid cathodes was further investigated [54, 55]. A barrier effect of PEDOT was demonstrated, which could prevent direct contact of the manganese oxide with the electrolyte and its degradation products. This effect might benefit the improvement of cycling stability of the lithium manganese oxide cathode. After electrochemical performance tests, the results indicated that the capacity fade of $LiMn_2O_4/PEDOT$ composite electrode was slightly lower than that of conventional $LiMn_2O_4/C$ cathode [54]. Although this attempt of introducing PEDOT into traditional cathode was not able to satisfy the cycle-life target of LIBs at that time, the series of studies still provided a promising direction and also gained meaningful results.

Additionally, Wallace's group insisted to improve the electrochemical performance of Li-based batteries with CPs cathodes during this timeframe [59-61]. In 2006, they developed polythiophene (PBiTh) as the cathode in lithium-polymer rechargeable batteries instead of the polypyrrole (PPy) electrode they fabricated before [59]. Due to more appropriate redox potentials, thiophene-based polymer electrodes could have better electrochemical/structural stability and electronic conductivity than that of PPy-based electrodes. According to this report, the as-prepared $PBiTh/PF_6$ coated stainless-steel mesh cathode presented a stable discharge capacity of *ca.* 82 mAh g^{-1} over 50 cycles (at 0.1 mA cm^{-2}), which was higher than that of PPy electrode. There was another important work reported in 2006, which opened another new gate for the CPs-based electrode materials. In consideration of the relationship between conductivity and morphology of polyaniline (PANI), Chen *et al.* synthesized PANI nanotubes and nanofibers and applied them as cathodes of Li-based batteries [65]. The percholoric acid-doped polyaniline (HClO4-PANI) nanotubes were fabricated by employing anodic aluminum oxide (AAO) membranes as templates, while the $HClO_4$-PANI nanotubes were prepared through a spray technique with steel mesh as the templates. When applied as the cathode of Li-based battery, the as-prepared $HClO_4$-PANI nanotubes exhibited larger charge/discharge capacities, better cycling stability as well as higher electrical conductivity than those of the commercial doped PNAI powders. This study not only provided a new strategy to improve electrochemical performance of CPs-based electrodes but also could be regarded as the prelude of nano-era of CPs-based electrodes.

2.3   Since 2007: Fast development stage

Up to 2007, the lithium-ion batteries have developed about 15 years and various cathode systems have been established, including $LiFePO_4$ and $Li_xM_yO_z$ (M = Mn, V, Co) [5-7].

Although some of these materials have been applied in the commercial Li-ion batteries, they still existed several problems which limited their electrochemical performance. Generally, these materials exhibited low ionic and electronic conductivities [66]. In order to overcome these defaults, one strategy was to reduce the particle size to decrease the length of the lithium-ion diffusion pathway [67]. Another way was to add conductive agent into electrode material to increase the electronic conductivity [68]. Considered these factors, conducting polymers were ideal candidates to fabricate hybrid electrodes combined with these conventional cathode materials. CPs could not only improve the electronic conductivity of the electrodes but also contribute certain specific capacities for the batteries [69]. Moreover, due to the flexibility and plasticity, CPs could easily combine with nanosized cathode materials via some surface modification techniques [70-72]. Therefore, since 2007, CPs-based nanosized hybrid cathodes gradually became the mainstream of CPs-based LIB cathodes. Meanwhile, the Li-polymer batteries gradually withdrew from the historical stage because of the flourish of Li-ion batteries. Generally, the CPs-based nanosized hybrid cathodes could be divided into two types during this period. One was nanoscale conventional cathode materials coated by conducting polymers [10, 69-89], another was conducting polymer framework on where nanoparticles of cathode materials loaded [90, 91].

At that time, $LiFePO_4$ was one of the most important and common cathode materials for commercial LIBs, thus a number of studies focused on how to improve the electrochemical performance of $LiFePO_4$ assisted with conductive polymers [71, 72, 75, 77, 78, 81]. In 2008, Manthiram et al. reported a novel microwave-solvothermal approach to fabricate well-defined nanoparticles of $LiFePO_4$ [71]. Then, they coated a mixed conducting polymer (p-toluene sulfonie acid-doped poly(3,4-ethylenedioxythiophene), denoted as p-TSA-PEDOT) on the $LiFePO_4$ nanoparticles to obtain nanohybrid cathode materials, as shown in Fig. 2(a). This nanohybrid electrode exhibited good cyclability with 3 % fade within 50 cycles, which was much better than those of pure $LiFePO_4$ and pure p-TSA PEDOT electrodes. The enhanced electrochemical performance of the hybrid cathode could be attributed to the mixed electronic-ionic conductivity of the p-TSA PEDOT. In 2011, Schougaard's team presented a new method to prepare $PEDOT/LiFePO_4$ composite electrode material, which could significantly improve the fabrication efficiency[78]. They selected the intrinsic oxidation power of $Li_{(1-x)}FePO_4$ instead of the external oxidant, which was the key point of this methodology. The $Li_{(1-x)}FePO_4$ acted as the driving force of the polymerization process. Firstly, the $Li_{(1-x)}FePO_4$ powders were synthesized by a delithiation process of $LiFePO_4$. Then, the polymerization of 3,4-ethylenedioxythiophene (EDOT) was realized by the reinsertion of lithium into $Li_{(1-x)}FePO_4$. The reaction equations are presented by Eq. 2 and Eq. 3.

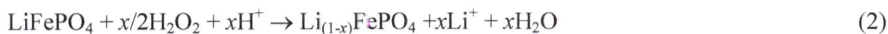

$$LiFePO_4 + x/2H_2O_2 + xH^+ \rightarrow Li_{(1-x)}FePO_4 + xLi^+ + xH_2O \tag{2}$$

$$Li_{(1-x)}FePO_4 + LiTFSI + EDOT \rightarrow PEDOT\text{-}LiFePO_4 \tag{3}$$

The as-fabricated PEDOT-LiFePO$_4$ composite cathode displayed good cycling stability within 30 cycles and also good rate performance.

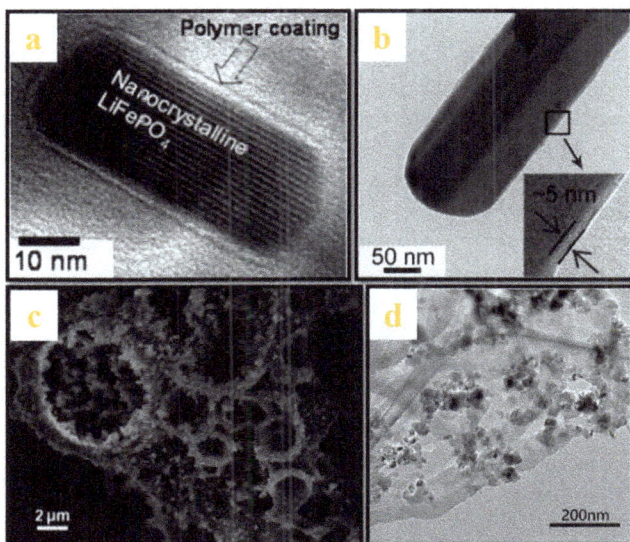

Figure 2   a) TEM image of the small LiFePO$_4$ nanorods after coating with p-TSA PEDOT, ref.[71], copyright 2008 Elsevier B.V.; b) TEM image of the V$_2$O$_5$@PANI core/shell nanofiber hybrids, ref.[19], copyright 2016 Elsevier Ltd.; c) SEM image of PEDOT/LiFePO$_4$ composite electrode, ref.[90], copyright 2012 Elsevier B.V.; d) TEM image of PPy/LFMO composite electrode, ref.[91], copyright of Springer-Verlag Berlin Heidelberg 2017.

Except for LiFePO$_4$, other kinds of cathode materials were also composited with conducting polymers to enhance their electrochemical performance. In 2014, Song and

Lithium-ion Batteries - Materials and Applications   Materials Research Forum LLC
Materials Research Foundations **80** (2020) 28-62   https://doi.org/10.21741/9781644900918-2

Lee synthesized LCO/PEDOT:PSS ($LiCoO_2$/poly(3,4-ethylenedioxythiophene) doped with poly(styrenesulfonate)) composite LIB cathode, where the PEDOT:PSS layers were coated on the LCO particles like skin[82]. The PEDOT:PSS skin could act as both binder and conductive agent, which was able to reach a higher energy density of LCO cathode. By adjusting the PSS content and also the structural conformation (benzoid-favoring and quinoid-favoring) of PEDOT, the electric conductivity of this composite electrode could be controlled, resulting in different electrochemical performance. In terms of electrochemical performance, the LCO@3.3Q-PEDOT:PSS electrode (3.3Q, representing the quinoid-type PEDOT:PSS with 3.3 mol of monomeric units of PSS per mole of PEDOT) exhibited the best results, which retained a reversible capacity over 400 mAh g^{-1} after 100 cycles at 1 C. In recent years, vanadium pentoxide (V_2O_5) has been widely studied as cathode material for lithium-ion batteries, due to ability of the vanadium ion to change its oxidation state (V-II) and low cost[92, 93]. In 2016, Li's team prepared V_2O_5@PANI core/shell nanofiber hybrids, which aimed to enhance the electronic conductivity, decrease the dissolution rate in liquid electrolytes and volume changes during ion intercalation/removal processes of the V_2O_5 cathode[10]. The V_2O_5 nanofibers were grown on fluorine-doped tin oxide (FTO) substrates by a controlled vertical drying method and anodic deposition process[94]. The PANI shells were then coated on V_2O_5 nanofibers via an electropolymerization process. The structure and morphology of as-prepared V_2O_5@PANI core/shell nanofiber hybrids are presented in Fig. 2(b). The thickness of the PANI shell was adjusted to 5 nm, 13 nm, and 32 nm by different deposition time of 30 s, 60 s and 240 s, respectively. With electrochemical investigations, it was indicated that the existence of PANI shell could effectively improve Li-ion storage performance of the V_2O_5-based cathode, which could be ascribed to the reduced interfacial electric resistance and inhibition of vanadium dissolution. Meanwhile, the electrochemical performance of hybrid cathode was dependent to the thickness of the PANI shell, that the deposition time. As a result, the sample V_2O_5@PANI-240s exhibited the best cycling stability.

Creating a conducting polymer framework for conventional cathode materials was another strategy to construct CPs-based hybrid cathodes for LIBs. In 2011, Wang *et al.* reported a novel method called dynamic solid/liquid (oil)/liquid (water) three-phase interline electropolymerization (D3PIE), which could be used to synthesize free-standing PEDOT thin films [95]. The oil phase (dichloromethane) was hydrophobic monomer (3,4-ethylenedioxythiophene or pyrrole), while the aqueous phase was only dopant electrolytes. The solid phase was a platinum wire which immersed vertically across the oil/water interface and acted as a working electrode. The polymer film grew on the interface between the oil and water phase when an overpotential was applied to the

Materials Research Forum LLC
https://doi.org/10.21741/9781644900918-2

electrode. As-prepared PEDOT film exhibited heterogeneous micro-nano-structure. Based on this study, in 2015, Schougaard *et al.* synthesized PEDOT/LiFePO$_4$ composite electrode by the D3PIE method, where LiFePO$_4$ particles dispersed in the PEDOT matrix, as presented in Fig. 2(c) [90]. The PEDOT framework not only provided mechanical support for LiFePO$_4$ particles but also enhanced the conductivity of the whole electrode, leading to improved electrochemical performance. In 2018, Zhao *et al.* prepared PPy/LFMO (polypyrrole/Li$_{1.26}$Fe$_{0.22}$Mn$_{0.52}$O$_2$) composite LIB cathode, where LFMO particles loaded on PPy nanowires, as shown in Fig. 2(d) [91]. The three-dimensional electrical conduction networks built by PPy nanowires could effectively enhance the conductivity and stability of the composite electrode. As a result, the PPy/LFMO composite electrode presented enhanced cycling capability, initial coulombic efficiency as well as rate performance. It reached an initial discharge capacity of 205 mAh g^{-1} at 1 C (1 C=200 mA g^{-1}), and kept reversible capacities of 132 mAh g^{-1} and 98 mAh g^{-1} after 50 cycles at 1 C and 3 C, respectively.

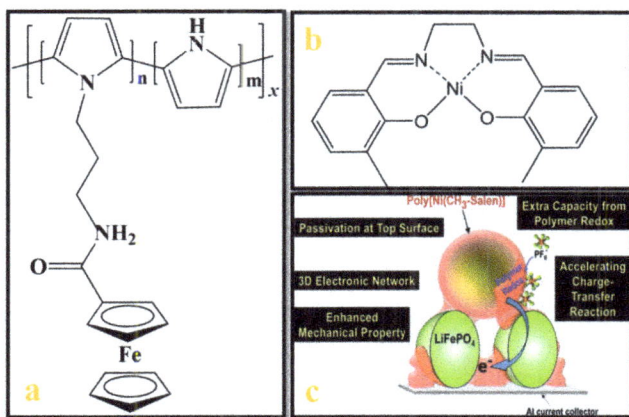

Figure 3   a) chemical structure of the pyrrole/[(ferrocene) amidopropyl]pyrrole copolymer, repainted according to ref.[96], copyright 2007 WILEY-VCH Verlag GmbH & Co. KGaA, Weinheim; b) chemical structure of Ni(CH$_3$-salen) monomer; c) multiple roles of poly[Ni(CH$_3$-salen)] in carbon-free cathodes, ref.[103], copyright 2018 American Chemical Society.

Since 2007, several new-type conducting polymers were developed and applied in the cathode materials for lithium-ion battery [96-103]. In 2007, Goodenough's team proposed a novel method about attaching a suitable redox couple to the polymer backbone, in order to improve the specific capacity of conducting polymers [96]. They anchored ferrocene groups to the PPy backbone to generate [(ferrocenyl) amidopropyl]pyrrole (denoted as PPy/ferrocene) conductive copolymer, which chemical structure is presented in Fig. 3(a). Incorporated this polymer with LiFePO$_4$, the specific capacity of this composite electrode was obviously increased. The PPy/LiFePO$_4$ cathode could display a specific capacity of over 130 mAh g^{-1} at 0.2 C, while the LiFePO$_4$/C/PTFE (75:20:5 in wt %) cathode exhibited less than 110 at the same condition. PEDOT has been studied as LIB cathode material for several years and gained much progress. In 2011, Abruna *et al.* synthesized some CPs which had a similar chemical structure to PEDOT but with different heteroatoms [98]. Those polymers were poly(3,4-ethylenedioxyselenophene) (PEDOS), and poly(3,4-ethylenedioxypyrrole) (PEDOP), corresponding to selenium and nitrogen, respectively. The as-prepared PEDOS and PEDOP exhibited good electrocatalytic activity to 2,5-dimercapto1,3,4-thiadiazole (DMcT) which is also a promising cathode material for LIBs [104-106]. Based on this study, PEDOX (X=Se, NH)/DMcT displayed a good potential to be high-performance cathode materials for Li-ion batteries. Recently, an interesting type of conducting polymer emerged and attracted attention as LIB electrode material, which known as the salen-type polymer [102, 103, 107-109]. Generally, the salen-type polymer can be synthesized by oxidative electrochemical polymerization of transition-metal complexes with salen-type N$_2$O$_2$ Schiff base ligands (salen is N,N'-bis(salicylidene)ethylenediamine), denoted as poly[M(Schiff)] (M is a transition metal, Schiff is a salen-type N$_2$O$_2$ Schiff base). In 2019, Kim *et al.* fabricated poly[Ni(CH$_3$-salen)] as an additive for conventional cathodes in lithium-ion batteries, whose monomer structure is presented in Fig. 3(b) [103]. According to this study, the poly[Ni(CH$_3$-salen)] played multiple positive roles in the composite cathode (poly[Ni(CH$_3$-salen)]/LiFePO$_4$). As shown in Fig. 3(c), the poly[Ni(CH$_3$-salen)] was able to i) form a 3D electronic network, ii) accelerate charge-transfer reaction of the cathode, iii) contribute extra specific capacity, iv) enhance the mechanical property of the electrode, v) passivate the active materials at the top surface. Therefore, the poly[Ni(CH$_3$-salen)]/LiFePO$_4$ cathode exhibited better electrochemical performance than that of the conventional carbon/LiFePO$_4$ cathode. Moreover, conducting polymers could separately act as binder or conductive agent instead of constructing composites with conventional cathode materials. This strategy also could improve the electrochemical performance of LIB cathodes, and attracted some attention since 2007 [9, 110-114].

3.   Applications on anode materials

3.1   Before 2010: Emergence stage

The origin of conducting polymers applied in anode materials could be dated back to the 1980s before the birth of the commercial lithium-ion batteries [39, 115]. However, to our best knowledge, there were only a few published reports about the applications of CPs on LIB anodes before 2010 [116-119]. In 2006, Wan et al. fabricated $SnCl_2$/PAN composite anode via a low-temperature pyrolysis process [118]. Generally, the polyacrylonitrile (PAN) is not a typical conducting polymer, but its electrical conductivity can be improved after thermal treatment due to the chains undergoing cyclization to form a conjugated-chain chemical structure [120, 121]. Based on this idea, the PAN could enhance the conductivity of Sn-based anode and also alleviate the volume expansion of Sn during charge and discharge processes. The as-prepared $SnCl_2$/PAN composite anode could remain a reversible capacity of 490 mAh g^{-1} after 50 cycles at a current density of 30 mA g^{-1}. In 2007, Wallace et al. combined conducting polymer with aligned carbon nanotube (ACNT) arrays to obtain PEDOT/ACNT composite anode material [119]. This composite anode remained a reversible capacity of 265 mAh g^{-1} after 50 cycles at 0.1 mA cm^{-2}, which was better than other CNT-based anodes at that time [122].

3.2   Since 2010: Rising stage

Since 2010, various anode materials gained considerable progress, including alloying-type anodes (Si, Sn, Ge, Sb) and conversion-type anodes (TiO_2, $Li_4Ti_5O_{12}$) [13, 14, 123]. Although these anodes exhibited several advantages compared with the conventional graphite-based anodes, such as higher theoretical capacity and better rate cyclability, they still suffered from some problems. For the alloying-type anodes, huge volumetric expansion in lithiation process leads to fast capacity fading [124]. In terms of conversion-type anodes, they usually have stable cyclability but their electrochemical performance is limited by the poor electrical conductivity [123]. Based on unique properties, conducting polymers were widely used in both alloying-type and conversion-type anodes in order to improve their electrochemical performance.

Due to good cycling stability and rate performance, titanium dioxide (TiO_2) has become one of the most favorite anodes for Li-ion batteries in recent years [125, 126]. However, it still has some drawbacks, such as relatively low specific capacity and electrical conductivity. In 2010, Dziewonski and Grzeszczu prepared TiO_2-PPy and TiO_2-PEDOT anodes where the conducting polymers could not only provide some specific capacity but also enhance the conductivity of the electrodes [127]. According to this report, the existence of PPy and PEDOT benefited the intercalation of Li ions into amorphous TiO_2

as well as the reversibility of the process, leading to better electrochemical performance of the composite anodes. Zinc oxide (ZnO), as a newly anode material, has attracted much attention due to its high theoretical capacity of 978 mAh g^{-1} [128-130]. In 2015, Chen and Amine introduced PEDOT:PSS into ZnO/C anode, in order to improve its conductivity and accommodate the volume change during cycling [131]. Assisted with PEDOT:PSS surface coating, the electrochemical performance of this anode was obviously enhanced. It maintained a reversible capacity of 748 mAh g^{-1} after 500 cycles at 0.5 C and 624 mAh g^{-1} after 1500 cycles at 1 C. Li$_4$Ti$_5$O$_{12}$ (LTO) has been successfully applied in commercial lithium-ion batteries due to its excellent cycling stability [132-134]. However, poor electron conductivity limits its rate performance [135]. In 2017, Xu *et al.* coated polythiophene (PTh) on the surface of LTO particles, to obtain core-shell structured LTO/PTh composite anode[136]. After PTh-coating, electron conductivities LTO anode increased from 7.3×10^{-9} S cm^{-1} (pure LTO) to 1.6×10^{-6} S cm^{-1} (LTO/PTh). Therefore, the LTO/PTh composite anode could reach a specific capacity of 157 mAh g^{-1} at 5 C, which was much better than that of pure LTO anode (130 mAh g^{-1}).

Silicon (Si) has been regarded as the most promising anode candidate for the next-generation lithium-ion batteries due to its highest theoretical capacity among all anode materials [14]. With the great advantage, the practical performance of Si and other alloying-type anode materials were still insufficient due to the huge volume expansion upon lithiation, as mentioned above. Therefore, since 2010, many efforts have done which aimed at using conducting polymers to improve the cycling stability of Si-based (or alloying-type) anodes. Similar to what discussed in the last part (2.3), the related studies about applications of CPs on alloying-type anodes could be divided into two classifications, one about coating CPs on active particles [137-146], another about loading active particles on CPs networks[147-149]. In 2012, Cui's team coated PEDOT on Si nanowires (NWs) to enhance the cyclic stability of electrode [138]. Si nanowires were fabricated by Au-catalyzed chemical vapor deposition, then PEDOT was coated on their surfaces via an electropolymerization process. The schematic morphology of as-prepared Si NW/PEDOT composite is displayed in Fig. 4(a). Finally, this composite anode retained a reversible capacity of 2510 mAh g^{-1} after 100 cycles at 0.2 C (1 C=4200 mA g^{-1}), corresponding to a capacity retention of 80 %. Compared to the bare Si NW anode, the capacity retention of Si NWs/PEDOT electrode increased from 30 % to 80 %. Up to 2013, Cui's team reported their another work about Si/CP hybrid anode for LIBs [139]. Si nanoparticles (Si NPs) were mixed phytic acid and aniline in water, where the phosphoric acid groups in the phytic acid molecules could bind with the SiO$_2$ on the Si particle surfaces via hydrogen bonding [150, 151]. Furthermore, the coated phytic acid molecules on the Si NPs could be crosslinked with aniline monomers during polymerization, forming a 3D conducting hydrogel network which trapped coated Si nanoparticles inside. As presented in Fig. 4(b), this study combined coating and loading strategies, resulting in enhanced electrochemical performance of the Si-based anode. This Si NP/PANI hybrid anode exhibited stable cyclability, which maintained a reversible capacity of 1600 mAh g^{-1} after 1000 cycles at 1.0 A g^{-1}. In addition, this hybrid anode also gained good rate performance, which could keep a stable capacity of 550 mAh g^{-1} after 5000 cycles at a high current density of 6.0 A g^{-1}. This study provided a novel and promising way to fabricate high-performance Si-based anodes by a scale-up manufacture method. In the following years, a variety of CPs-based composite LIB anodes were synthesized and gained enhanced electrochemical performance, including Sn/PFM (Poly(9,9-dioctylfluorene-co-fluorenone-comethylbenzoicester)) [140], Si/PEDOT:PSS [141, 142], Si/SiO$_x$-PEDOT:PSS [143], Si/C-PPy [144], Si/EP(EDOT-phenylene-based conjugated polymer) [146] composite materials.

In terms of the strategy about building CPs networks for active materials, it also gained some good results [147-149]. In 2013, Yu *et al.* prepared a 3D hierarchically porous framework established by polypyrrole (PPy) and single-walled carbon nanotubes (SWCNTs), to act as the matrix for Si nanoparticles (Si NPs) [147]. As presented by Fig. 4(c), the 3D conductive framework was formed by the nanostructured PPy hydrogel. Meanwhile, a thin *in-situ* polymeric coating layer was generated on the surface of Si NPs, which could further link with the external conductive polymeric framework. Such a unique design could connect the separated Si nanoparticles with the conducting network, forming an interconnected and stable nanostructure. Such structure was beneficial to accommodate the huge volumetric changes of silicon and improved the electrical conductivity of the whole electrode. Moreover, the addition of SWCNTs could further enhance the stability and conductivity of the electrode, due to the roles of the wrapping layer and conductive backbone. Therefore, good electrochemical performance of this Si/PPy/CNT composite anode was expected. As a result, this anode achieved a reversible capacity of *ca.* 1600 mAh g^{-1} after 1000 cycles at a current density of 3.3 A g^{-1}, corresponding to a capacity retention of over 85 %. When tested at a higher current density of 8.6 A g^{-1}, the Si/PPy/CNT composite anode still exhibited very stable cyclability, which retained a specific capacity of 1000 mAh g^{-1} after 2000 cycles. In 2014, Bao and Cui applied a similar strategy to synthesize flexible hybrid anodes [148]. PEDOT:PSS, carbon nanotubes (CNTs) and active nanoparticles (including TiO$_2$ and Si) were used to fabricate 3D interconnected carbon nanotube-conducting polymer hydrogel network for flexible battery electrode, as displayed in Fig. 4(d). The as-fabricated TiO$_2$/PEDOT:PSS/CNT composite anode displayed great flexibility. Its resistance only presented some tiny changes after 500 bending cycles at a bending radius of 3.5 mm, indicating its good overall mechanical property. Besides, due to enhanced electrical conductivity, the TiO_2/PEDOT:PSS/CNT anode also exhibited good rate performance. In case of the Si/PEDOT:PSS/CNT electrode, it could keep about 90 % of its initial capacity after 100 cycles, while a control electrode made by mixing CNT with PEDOT:PSS coated SiNP lost about 55% of its capacity.

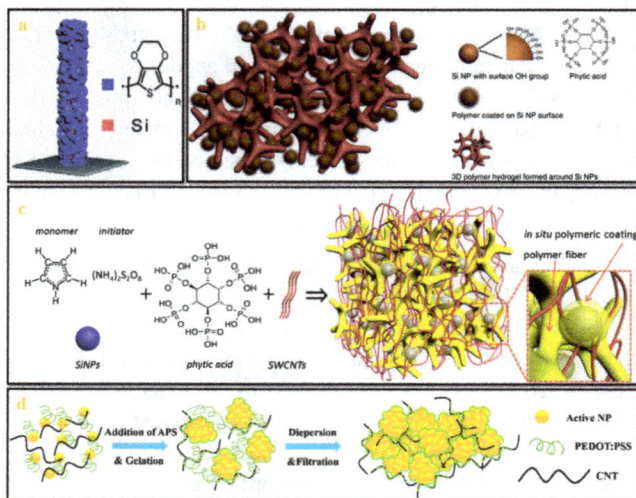

Figure 4   a) Schematic of a PEDOT (blue)-coated Si NW(brown) with the molecular structure of PEDOT shown, ref.[138], copyright The Royal Society of Chemistry 2012; b) Schematic illustration of 3D porous SiNP/conductive polymer hydrogel composite electrodes, ref.[139], copyright 2013 Macmillan Publishers Limited.; c) Schematic illustration of the formation of 3D Si/PPy/CNT ternary electrode, ref.[147], copyright 2013 American Chemical Society; d) Schematic of the aqueous solution process to fabricate flexible electrodes using active nanoparticles (e.g., TiO_2 or Si nanoparticles), CNT and PEDOT:PSS, repainted according to ref.[148], copyright 2014 WILEY-VCH Verlag GmbH & Co. KGaA, Weinheim.

In recent years, several novel conducting polymers gained increasing popularity as anodes for lithium-ion batteries, especially the conjugated ladder polymers (CLPs) [152-157]. Ladder or ribbon polymers consist of cyclic subunits, connected to each other by two links which are attached to different sites of the respective subunits [158]. The history of synthesizing ladder polymers could be dated back to 1970s when the poly(benzimidazobenzophenanthroline) (BBL) was synthesized by the U.S. Air Force

Wright Aeronautical Laboratory [159]. The conjugated ladder polymers are those ladder polymers with π-conjugated structures, which exhibit good electrical conductivity and have been applied in various electronic devices [160-164]. With large π-conjugated structure, rich nitrogen heteroatoms and multiring aromatics, the conjugated ladder polymers are also promising anode materials for Li-ion batteris [152]. From 2015 to 2016, Zhang and Yan led their team to reported a series of studies about CLPs-based anode materials for lithium-ion batteries. In 2015, they synthesized polyazaacene analogue poly(1,6-dihydropyrazino[2,3g]quinoxaline-2,3,8triyl-7-(2H)-ylidene-7,8dimethylidene) (PQL) and demonstrated its good lithium storage property [152]. As shown in Fig. 5(a), the PQL was obtained by the reaction between 2,5-dihydroxy-1,4-benzoquinone and 1,2,4,5-tetraaminobenzene tetrahydrobromide (TAB·4HBr) in 116% polyphosphoric acid (PPA) at 190 °C for 10 h [165]. Applied as anode for LIBs, the PQL could retain a reversible capacity of 1550 mAh g^{-1} after 100 cycles at a current density of 100 mA g^{-1}. Moreover, the PQL exhibited an electrical conductivity of $2.1×10^{-3}$ S cm^{-1}, indicating a good rate performance. Tested at 50°C, the PQL anode delivered a reversible capacity of about 500 mAh g^{-1} after 1000 cycles at a current density of 4555 mA g^{-1}. In 2015, they also reported their another work about CLPs-based anodes [153]. They synthesized BBL and its derivative, SBBL through one-pot polycondensation between 1,2,4,5-tetraaminobenzene tetrahydrobromide (TAB·4HBr) and pyromellitic dianhydride (PMDA)/naphthalenete-tracarboxylic dianhydride (NTCDA) in polyphosphoric acid (PPA), as displayed in Fig. 5(b) [165, 166]. In terms of electrochemical performance, the BBL electrode retained a reversible capacity of 496 mAh g^{-1} after 1000 cycles at 3 C rate (1 C=1962 mA g^{-1}), while the SBBL gained a capacity of 320 mAh g^{-1} at same test conditions (here 1 C=1888 mA g^{-1}). Compare to the SBBL, the enhanced electrochemical performance of BBL anode could be attributed to its part network or nonlinear structure in the polymer chain. Up to 2016, they fabricated another conjugated ladder polymer, called poly(1,4-dihydro11H-pyrazino[2',3':3,4]cyclopenta[1,2-b]quinoxalin-11-one) (PPCQ). Similar with their previous methods, the PPCQ was prepared by a simple one-step polymerization reaction, as presented in Fig. 5(b). This novel conjugated ladder polymer anode also exhibited good performance in both cycling and rate tests. The efforts made by Zhang and Yan's team not only extended the application range of conjugated ladder polymers but also provided a kind of promising anode material for lithium-ion batteries.

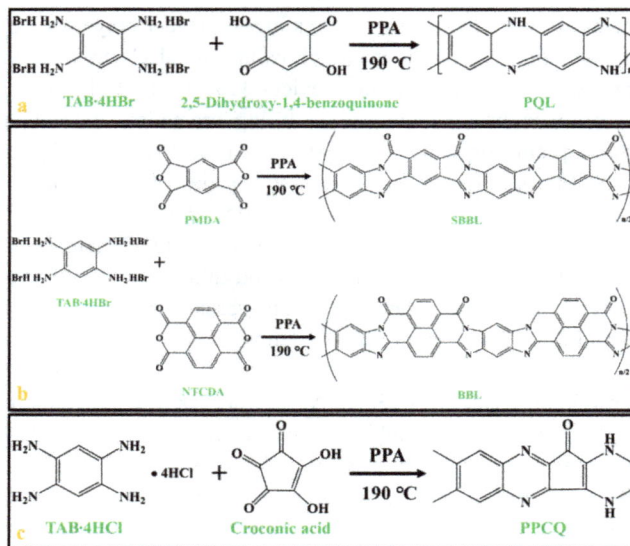

Figure 5   a) Synthesis of the ladder polymer PQL, repainted according to ref.[152], copyright 2015 Wiley-VCH Verlag GmbH &Co. KGaA, Weinheim; b) Synthetic scheme for the polymers of SBBL and BBL, repainted according to ref.[153], copyright 2015 WILEY-VCH Verlag GmbH & Co. KGaA, Weinheim; c) Synthetic route to the ladder-structured polymer PPCQ, repainted according to ref.[154], copyright 2016 American Chemical Society.

Conclusions & Outlooks

Conducting polymers were synthesized in 1976 for the first time, then attracted considerable attention in short order due to their unique electrical and physical properties. When lithium-ion batteries were successfully commercialized in 1992, conducting polymers were quickly applied as electrode materials. After about 30-years development, conducting polymers have become one of the most significant branches of electrode materials for lithium-ion batteries. A verity of conducting polymers have been developed and applied in both cathode and anode materials for Li-ion batteries, including PA, PANI, PPy, PTh, PEDOT, *et al.*

Applying conducting polymers in cathodes was always a research hotspot since the discovery of conductive polymer, whose origin could be dated back to 1980s. During the emergence stage before 2000, most of the studies focused on using pure CPs as cathodes

Lithium-ion Batteries - Materials and Applications   Materials Research Forum LLC
Materials Research Foundations **80** (2020) 28-62   https://doi.org/10.21741/9781644900918-2

for Li metal-based rechargeable batteries. From 2000 to 2006, conducting polymers started to combine with the normal cathode materials to improve electrochemical performance. Meanwhile, there was still a part of studies which aimed to build lithium-polymer batteries assisted with conductive polymers. Since 2007, lithium-ion batteries have gained fruitful achievements and played important roles as portable power sources in daily life. The flourish of cathode materials also exposed several problems which needed to be solved as soon as possible, such as low ionic and electronic conductivities. Nanotechnology gained rocket-like development in the new century and has been utilized to overcome the problems of LIB cathodes. Therefore, the combination of conducting polymer and nanotechnology became a new mainstream to fabricate high-performance composite cathodes during this period. Two different strategies for designing the CPs-based composite cathodes appeared. One was coating the conducting polymers on nanoparticles of the conventional cathode material, where PEDOT:PSS and LiFePO$_4$ attracted much attention. Another strategy was to creating conducting polymer framework to warp the nanoparticles. Due to more facile synthesis, which usually realized by an electropolymerization process, the first trend exhibited a dominated position among all related studies. Moreover, in recent years, several new-type conducting polymers were synthesized and displayed the great potential to act as high-performance LIB cathodes, including some analogs of PEDOT and salen-type polymers.

In terms of CPs-based anodes, the origin could be de dated back to the 1980s. However, compared to the cathode materials, there were only a few studies about CPs-based anodes for LIB before 2010. This could be attributed to the great success of carbon-based anodes in the first ten years of the 21st century. Since 2010, the limited specific capacity of graphite-based anodes promoted the wave of looking for high-capacity anode materials. Various newly anode materials were synthesized and studied, including alloying-type and conversion-type anodes. Nevertheless, most of these anodes suffered from some drawbacks, such as the volumetric effect of alloying-type anodes and low conductivity of conversion-type anodes. Therefore, conducting polymers embraced their golden period of applications in LIB anodes due to the good electrical and mechanical characteristics, which were beneficial to solve those problems above. Similar to the cathode, the studies about CPs-based composite anodes could also be divided into two types, coating, and loading design strategies. Aiming to improve cycling stability of Si anodes, the most promising anode for the future LIBs, various conducting polymer/Si nanostructure composite materials were fabricated and exhibited very competitive electrochemical performance. In addition, assisted with conducting polymer, some conversion-type anodes, such as TiO$_2$ and Li$_4$Ti$_5$O$_{12}$, also displayed enhanced electrochemical performance. In the last several years, the conjugated ladder polymers emerged as novel anodes and also exhibited stable cyclability with high reversible capacity.

Due to the combined characteristics of conductor and polymer in the conducting polymers, they could be regarded as ideal additives or modification agents for both cathodes and anodes of Li-ion batteries. Conducting polymers can easily composite with various conventional electrode materials because of the good plasticity, to overcome some problems of these materials and enhance their electrochemical performance. In many studies, the conducting polymer can act as both binder and conductive agent, which benefits to increase the energy density of batteries. Moreover, the conducting polymers can endow the flexibility to the electrode, thus exhibit a broad application prospect for flexible and wearable electron devices in the future. According to Alan J. Heeger's classification, conducting polymers belong to the fourth generation of polymer materials[27]. It is believed that this fourth-generation polymer will play a significant role to promote the development of second-generation lithium-ion batteries.

Acknowledgment

This research is funded by the National Key R&D Program of China (Grant No. 2016YFB0100100), Natural Science Foundation of China (51702335, 21773279), the CAS-EU S&T cooperation partner program (174433KYSB20150013), and Key Laboratory of Bio-based Polymeric Materials of Zhejiang Province. S.L. acknowledges the China Scholarship Council (CSC). P.M-B acknowledges funding by the International Research Training Group 2022 Alberta/Technical University of Munich International Graduate School for Environmentally Responsible Functional Hybrid Materials (ATUMS).

References

[1] T. Nagaura, K. Tozawa, Lithium ion rechargeable battery, Prog. Batteries Solar Cells 9 (1990) 79-104.

[2] B. Scrosati, Lithium rocking chair batteries: an old concept?, J. Electrochem. Soc. 139 (1992) 2776-2781. https://doi.org/10.1149/1.2068978

[3] A. Yoshino, The Birth of the Lithium B9 (1992) 2776-278. Chem. Int. Ed. 51 (2012) 5798-5800. https://doi.org/10.1002/anie.201105006

[4] M. Winter, B. Barnett, K. Xu, Before Li ion batteries, Chem. Rev. 118 (2018) 11433-11456. https://doi.org/10.1021/acs.chemrev.8b00422

[5] B. Dunn, H. Kamath, J.M. Tarascon, Electrical energy storage for the grid: A battery of choices, Science 334 (2011) 928-935. https://doi.org/10.1126/science.1212741

[6] V. Etacheri, R. Marom, R. Elazari, G. Salitra, D. Aurbach, Challenges in the development of advanced Li-ion batteries: a review, Energy Environ. Sci. 4 (2011) 3243-3262. https://doi.org/10.1039/c1ee01598b

[7] J.M. Tarascon, M. Armand, Issues and challenges facing rechargeable lithium batteries, Materials for sustainable energy: A collection of peer-reviewed research and review articles from nature publishing group, World Scientific2011, pp. 171-179. https://doi.org/10.1142/9789814317665_0024

[8] M.V. Reddy, G.V. Subba Rao, B.V. Chowdari, Metal oxides and oxysalts as anode materials for Li ion batteries, Chem. Rev. 113 (2013) 5364-5457. https://doi.org/10.1021/cr3001884

[9] H.D. Lee, G.J. Jung, H.S. Lee, T. Kim, J.D. Byun, K.S. Suh, Improved Stability of Lithium-Ion Battery Cathodes Using Conducting Polymer Binders, Sci. Adv. Mater. 8 (2016) 84-88. https://doi.org/10.1166/sam.2016.2606

[10] Z.Q. Tong, S.K. Liu, X.G. Li, Y.B. Ding, J.P. Zhao, Y. Li, Facile and controllable construction of vanadium pentoxide@conducting polymer core/shell nanostructures and their thickness-dependent synergistic energy storage properties, Electrochimi. Acta 222 (2016) 194-202. https://doi.org/10.1016/j.electacta.2016.09.098

[11] B. Xiao, X. Sun, Surface and subsurface reactions of lithium transition metal oxide cathode materials: An overview of the fundamental origins and remedying approaches, Adv. Energy Mater. 8 (2018) 1802057. https://doi.org/10.1002/aenm.201802057

[12] M.N. Obrovac, V.L. Chevrier, Alloy negative electrodes for Li-ion batteries, Chem. Rev. 114 (2014) 11444-11502. https://doi.org/10.1021/cr500207g

[13] D. Liu, Z.J. Liu, X. Li, W. Xie, Q. Wang, Q. Liu, Y. Fu, D. He, Group IVA element (Si, Ge, Sn)-based alloying/dealloying anodes as negative electrodes for full-cell lithium-ion batteries, Small 13 (2017) 1702000. https://doi.org/10.1002/smll.201702000

[14] X. Zuo, J. Zhu, P. Müller-Buschbaum, Y.-J. Cheng, Silicon based lithium-ion battery anodes: A chronicle perspective review, Nano Energy 31 (2017) 113-143. https://doi.org/10.1016/j.nanoen.2016.11.013

[15] D. Chen, Z. Lou, K. Jiang, G.Z. Shen, Device configurations and future prospects of flexible/stretchable lithium-ion batteries, Adv. Funct. Mater. 28 (2018) 1805596. https://doi.org/10.1002/adfm.201805596

[16] Q.J. Huang, Y. Zhu, Printing Conductive Nanomaterials for Flexible and Stretchable Electronics: A Review of Materials, Processes, and Applications, Adv. Mater. Technol. 4 (2019) 1800546. https://doi.org/10.1002/admt.201800546

[17] Y.N. Jia, X.X. Liu, Y. Lu, Y.F. Su, R.J. Chen, F. Wu, Flexible electrode assembled from different microstructures, Progress Chem. 31 (2019) 464-474.

[18] C.Y. Wang, K.L. Xia, H.M. Wang, X.P. Liang, Z. Yin, Y.Y. Zhang, Advanced carbon for flexible and wearable electronics, Adv. Mater. 31 (2019) 1801072. https://doi.org/10.1002/adma.201801072

[19] M.E. Abdelhamid, A.P. O'Mullane, G.A. Snook, Storing energy in plastics: A review on conducting polymers & their role in electrochemical energy storage, RSC Adv. 5 (2015) 11611-11626. https://doi.org/10.1039/C4RA15947K

[20] P. Sengodu, A.D. Deshmukh, Conducting polymers and their inorganic composites for advanced Li-ion batteries: A review, RSC Adv. 5 (2015) 42109-42130. https://doi.org/10.1039/C4RA17254J

[21] P.T. Xiao, F.X. Bu, G.H. Yang, Y. Zhang, Y.X. Xu, Integration of graphene, nano sulfur, and conducting polymer into compact, flexible lithium-sulfur battery cathodes with ultrahigh volumetric capacity and superior cycling stability for foldable devices, Adv. Mater. 29 (2017) 1073324. https://doi.org/10.1002/adma.201703324

[22] Z. Song, H. Zhou, Towards sustainable and versatile energy storage devices: an overview of organic electrode materials, Energy Environ. Sci. 6 (2013) 2280-2301. https://doi.org/10.1039/c3ee40709h

[23] S. Lee, G. Kwon, K. Ku, K. Yoon, S.K. Jung, H.D. Lim, K. Kang, Recent progress in organic electrodes for Li and Na rechargeable batteries, Adv. Mater. 30 (2018) 1704682. https://doi.org/10.1002/adma.201704682

[24] G. Milczarek, O. Inganäs, Renewable cathode materials from biopolymer/conjugated polymer interpenetrating networks, Science 335 (2012) 1468-1471. https://doi.org/10.1126/science.1215159

[25] J. Heinze, Electrochemistry of conducting polymers, Synth. Met. 43 (1991) 2805-2823. https://doi.org/10.1016/0379-6779(91)91183-B

[26] X.C. Li, Y.S. Jiao, S.J. Li, The synthesis, properties and application of new conducting polymers, Eur. Polym. J. 27 (1991) 1345-1351. https://doi.org/10.1016/0014-3057(91)90233-E

[27] A.J. Heeger, Semiconducting and metallic polymers: the fourth generation of polymeric materials, ACS Publications, 2001. https://doi.org/10.1557/mrs2001.232

[28] H. Shirakawa, E.J. Louis, A.G. MacDiarmid, C.K. Chiang, A.J. Heeger, Synthesis of electrically conducting organic polymers: Halogen derivatives of polyacetylene,(CH)x, J. Chem. Soc., Chem. Commun. (1977) 578-580. https://doi.org/10.1039/c39770000578

[29] C. Chiang, M. Druy, S. Gau, A. Heeger, E. Louis, A.G. MacDiarmid, Y. Park, H. Shirakawa, Synthesis of highly conducting films of derivatives of polyacetylene,(CH)$_x$, J. Am. Chem. Soc. 100 (1978) 1013-1015. https://doi.org/10.1021/ja00471a081

[30] Applications of Conducting Polymers, Conducting Polymers: A New Era in Electrochemistry 2008, pp. 225-263.

[31] L.J. Pan, H. Qiu, C.M. Dou, Y. Li, L. Pu, J.B. Xu, Y. Shi, Conducting Polymer Nanostructures: Template Synthesis and Applications in Energy Storage, Int. J. Mol. Sci. 11 (2010) 2636-2657. https://doi.org/10.3390/ijms11072636

[32] X.F. Lu, W.J. Zhang, C. Wang, T.C. Wen, Y. Wei, One-dimensional conducting polymer nanocomposites: Synthesis, properties and applications, Prog. Polym. Sci. 36 (2011) 671-712. https://doi.org/10.1016/j.progpolymsci.2010.07.010

[33] T.K. Das, S. Prusty, Review on Conducting Polymers and Their Applications, Polym. Plast. Technol. 51 (2012) 1487-1500. https://doi.org/10.1080/03602559.2012.710697

[34] Y. Shi, L.L. Peng, G.H. Yu, Nanostructured conducting polymer hydrogels for energy storage applications, Nanoscale 7 (2015) 12796-12806. https://doi.org/10.1039/C5NR03403E

[35] M.H. Naveen, N.G. Gurudatt, Y.B. Shim, Applications of conducting polymer composites to electrochemical sensors: A review, App. Mater. Today 9 (2017) 419-433. https://doi.org/10.1016/j.apmt.2017.09.001

[36] M.E. Bhosale, S. Chae, J.M. Kim, J.Y. Choi, Organic small molecules and polymers as an electrode material for rechargeable lithium ion batteries, J. Mater. Chem. A 6 (2018) 19885-19911. https://doi.org/10.1039/C8TA04906H

[37] S. Ghosh, T. Maiyalagan, R.N. Basu, Nanostructured conducting polymers for energy applications: towards a sustainable platform, Nanoscale 8 (2016) 6921-6947. https://doi.org/10.1039/C5NR08803H

[38] P.J. Nigrey, D. MacInnes, D.P. Nairns, A.G. MacDiarmid, A.J. Heeger, Lightweight Rechargeable Storage Batteries Using Polyacetylene,(CH)x as the Cathode-Active Material, J. Electrochem. Soc. 128 (1981) 1651-1654. https://doi.org/10.1149/1.2127704

[39] K. Kaneto, K. Yoshino, Y. Inuishi, Characteristics of polythiophene battery, JPN. J. App. Phys. 22 (1983) L567-L572. https://doi.org/10.1143/JJAP.22.L567

[40] N. Mermilliod, J. Tanguy, F. Petiot, A study of chemically synthesized polypyrrole as electrode material for battery applications, J. Electrochem. Soc. 133 (1986) 1073-1079. https://doi.org/10.1149/1.2108788

[41] R. Bittihn, G. Ely, F. Woeffler, H. Münstedt, H. Naarmann, D. Naegele, Polypyrrole as an electrode material for secondary lithium cells, Makromolekulare Chemie. Macromolecular Symposia, Wiley Online Library, 1987, pp. 51-59. https://doi.org/10.1002/masy.19870080106

[42] L.S. Yang, Z.Q. Shan, Y.D. Liu, Performance of polyaniline positive in a lithium battery, J. Power Sources 34 (1991) 141-145. https://doi.org/10.1016/0378-7753(91)85033-S

[43] A.P. Chattaraj, I.N. Basumallick, Improved conducting polymer cathodes for lithium batteries, J. Power Sources 45 (1993) 237-242. https://doi.org/10.1016/0378-7753(93)87013-S

[44] T. Osaka, T. Momma, Impedance analysis of electropolymerized conducting polymers for polymer battery cathodes, Electrochim. Acta 38 (1993) 2011-2014. https://doi.org/10.1016/0013-4686(93)80333-U

[45] T. Osaka, T. Momma, K. Shiota, S. Nakamura, Electrochemical evaluation of a polyanline/polupyrrole dual-layer for rechargeable lithium battery cathode, Denki Kagaku 61 (1993) 1361-1365. https://doi.org/10.5796/electrochemistry.61.1361

[46] T. Boinowitz, G. tom Suden, U. Tormin, H. Krohn, F. Beck, A metal-free polypyrrole/graphite secondary battery with an anion shuttle mechanism, J. Power Sources 56 (1995) 179-187. https://doi.org/10.1016/0378-7753(95)80031-B

[47] M.K. Song, W.I. Jung, H.W. Rhee, Flexible Polymer Battery with Conducting Polymer As a Cathode, Mol. Cryst. Liq. Cryst. Sci. Technol., Sect. A, Mol. Cryst. Liq. Cryst. 316 (1998) 337-340. https://doi.org/10.1080/10587259808044523

[48] S. Kuwabata, T. Idzu, S. Masui, H. Yoneyama, Use of conducting polymers as conducting matrix and binder for preparation of positive electrodes of metal oxide particles in lithium secondary batteries, 1999.

[49] C. Arbizzani, M. Mastragostino, M. Rossi, Preparation and electrochemical characterization of a polymer $Li_{1.03}Mn_{1.97}O_4$/pEDOT composite electrode, Electrochem. Commun. 4 (2002) 545-549. https://doi.org/10.1016/S1388-2481(02)00368-5

[50] A.K. Cuentas-Gallegos, R. Vijayaraghavan, M. Lira-Cantu, N. Casan-Pastor, P. Gomez-Romero, Hybrid materials based on vanadyl phosphate and conducting

polymers as cathodes in rechargeable lithium batteries, Bol. Soc. Esp. Ceram. V. 43 (2004) 429-433. https://doi.org/10.3989/cyv.2004.v43.i2.545

[51] M. Bengoechea, I. Boyano, O. Miguel, I. Cantero, E. Ochoteco, J. Pomposo, H. Grande, Chemical reduction method for industrial application of undoped polypyrrole electrodes in lithium-ion batteries, J. Power Sources 160 (2006) 585-591. https://doi.org/10.1016/j.jpowsour.2006.01.051

[52] H. Tsutsumi, H. Higashiyama, K. Onimura, T. Oishi, Preparation of poly (N-methylpyrrole) modified with pentathiepin rings and its application to pasitive active material for lithium secondary. J. Power Sources 146 (2005) 345-348. https://doi.org/10.1016/j.jpowsour.2005.03.015

[53] J.M. Pope, T. Sato, E. Shoji, N. Oyama, K.C. White, D.A. Buttry, Organosulfur/conducting polymer composite cathodes-II. Spectroscopic determination of the protonation and oxidation states of 2,5-dimercapto-1,3,4-thiadiazole, J. Electrochem. Soc. 149 (2002) A939-A952. https://doi.org/10.1149/1.1482768

[54] C. Arbizzani, A. Balducci, M. Mastragostino, M. Rossi, F. Soavi, Li1.01Mn1.97O4 surface modification by poly (3,4-ethylenedioxythiophene), J. Power Sources 119 (2003) 695-700. https://doi.org/10.1016/S0378-7753(03)00228-3

[55] C. Arbizzani, A. Balducci, M. Mastragostino, M. Rossi, F. Soavi, Characterization and electrochemical performance of Li-rich manganese oxide spinel/poly (3, 4-ethylenedioxythiophene) as the positive electrode for lithium-ion batteries, J. Electroanal. Chem. 553 (2003) 125-133. https://doi.org/10.1016/S0022-0728(03)00305-X

[56] E.C. Venancio, A.J. Motheo, F.A. Amaral, N. Bocchi, Performance of polyaniline electrosynthesized in the presence of trichloroacetic acid as a battery cathode, J. Power Sources 94 (2001) 36-39. https://doi.org/10.1016/S0378-7753(00)00659-5

[57] K.S. Ryu, Y. Lee, K.S. Han, M.G. Kim, The electrochemical performance of polythiophene synthesized by chemical method as the polymer battery electrode, Mater. Chem. Phys. 84 (2004) 380-384. https://doi.org/10.1016/j.matchemphys.2003.12.009

[58] C.C. Chang, L.J. Her, J.L. Hong, Copolymer from electropolymerization of thiophene and 3,4-ethylenedioxythiophene and its use as cathode for lithium ion battery, Electrochim. Acta 50 (2005) 4461-4468. https://doi.org/10.1016/j.electacta.2005.02.008

[59] J. Chen, J. Wang, C. Wang, C. Too, G. Wallace, Lithium–Polymer battery based on polybithiophene as cathode material, J. Power Sources 159 (2006) 708-711. https://doi.org/10.1016/j.jpowsour.2005.10.100

Materials Research Forum LLC
https://doi.org/10.21741/9781644900918-2

[60] J. Wang, C.O. Too, G.G. Wallace, A highly flexible polymer fibre battery, J. Power Sources 150 (2005) 223-228. https://doi.org/10.1016/j.jpowsour.2005.01.046

[61] J. Wang, C.O. Too, D. Zhou, G.G. Wallace, Novel electrode substrates for rechargeable lithium/polypyrrole batteries, J. Power Sources 140 (2005) 162-167. https://doi.org/10.1016/j.jpowsour.2004.08.040

[62] G.G. Amatucci, N. Pereira, T. Zheng, J.M. Tarascon, Failure Mechanism and Improvement of the Elevated Temperature Cycling of $LiMn_2O_4$ Compounds Through the Use of the $LiAl_xMn_{2-x}O_{4-z}F_z$ Solid Solution, J. Electrochem. Soc. 148 (2001) A171-A182. https://doi.org/10.1149/1.1342168

[63] D.D. MacNeil, T. Hatchard, J.R. Dahn, A Comparison Between the High Temperature Electrode/Electrolyte Reactions of Li_xCoO_2 and $Li_xMn_2O_4$, J. Electrochem. Soc. 148 (2001) A663-A667. https://doi.org/10.1149/1.1375798

[64] Y. Matsuo, R. Kostecki, F. McLarnon, Surface Layer Formation on Thin-Film $LiMn_2O_4$ Electrodes at Elevated Temperatures, J. Electrochem. Soc.148 (2001) A687-A692. https://doi.org/10.1149/1.1373658

[65] F.Y. Cheng, W. Tang, C.S. Li, J. Chen, H.K. Liu, P.W. Shen, S.X. Dou, Conducting poly(aniline) nanotubes and nanofibers: Controlled synthesis and application in lithium/poly(aniline) rechargeable batteries, Chem-Eur. J. 12 (2006) 3082-3088. https://doi.org/10.1002/chem.200500883

[66] M. Park, X. Zhang, M. Chung, G.B. Less, A.M. Sastry, A review of conduction phenomena in Li-ion batteries, J. Power Sources 195 (2010) 7904-7929. https://doi.org/10.1016/j.jpowsour.2010.06.060

[67] P.G. Bruce, B. Scrosati, J.M. Tarascon, Nanomaterials for rechargeable lithium batteries, Angew. Chem. Int. Ed. 47 (2008) 2930-2946. https://doi.org/10.1002/anie.200702505

[68] Y.H. Huang, K.S. Park, J.B. Goodenough, Improving lithium batteries by tethering carbon-coated $LiFePO_4$ to polypyrrole, J. Electrochem. Soc. 153 (2006) A2282-A2286. https://doi.org/10.1149/1.2360769

[69] S.Y. Chew, C. Feng, S.H. Ng, J. Wang, Z. Guo, H. Liu, Low-temperature synthesis of polypyrrole-coated LiV_3O_8 composite with enhanced electrochemical properties, J. Electrochem. Soc. 154 (2007) A633-A637. https://doi.org/10.1149/1.2734778

[70] C.V.S. Reddy, J. Wei, Z. Quan-Yao, D. Zhi-Rong, C. Wen, S. Mho, R.R. Kalluru, Cathodic performance of (V_2O_5+PEG) nanobelts for Li ion rechargeable battery, J. Power Sources 166 (2007) 244-249. https://doi.org/10.1016/j.jpowsour.2007.01.010

[71] A.V. Murugan, T. Muraliganth, A. Manthiram, Rapid microwave-solvothermal synthesis of phospho-olivine nanorods and their coating with a mixed conducting

polymer for lithium ion batteries, Electrochem. Commun. 10 (2008) 903-906. https://doi.org/10.1016/j.elecom.2008.04.004

[72] H.C. Dinh, I.H. Yeo, W.I. Cho, S.I. Mho, Characteristics of Conducting Polymer-Coated Nanosized LiFePO$_4$ Cathode in the Li+ Batteries, in: X. Zhang, D. Chu, P.S. Fedkiw, C. Wang (Eds.) Adv. Org. Inorg. Mater. Electrochem. Power Sources 2010, pp. 167-175. https://doi.org/10.1149/1.3490696

[73] G.C. Li, C.Q. Zhang, H.R. Peng, K.Z. Chen, One-Dimensional V$_2$O$_5$@Polyaniline Core/Shell Nanobelts Synthesized by an In situ Polymerization Method, Macromol. Rapid Comm. 30 (2009) 1841-1845. https://doi.org/10.1002/marc.200900322

[74] S.R. Sivakkumar, P.C. Howlett, B. Winther-Jensen, M. Forsyth, D.R. MacFarlane, Polyterthiophene/CNT composite as a cathode material for lithium batteries employing an ionic liquid electrolyte, Electrochim. Acta 54 (2009) 6844-6849. https://doi.org/10.1016/j.electacta.2009.06.091

[75] A. Fedorková, A. Nacher-Alejos, P. Gómez-Romero, R. Oriňáková, D. Kaniansky, Structural and electrochemical studies of PPy/PEG-LiFePO$_4$ cathode material for Li-ion batteries, Electrochim. Acta 55 (2010) 943-947. https://doi.org/10.1016/j.electacta.2009.09.060

[76] A. Fedorková, R. Oriňáková, A. Oriňák, H.-D. Wiemhöfer, D. Kaniansky, M. Winter, Surface treatment of LiFePO$_4$ cathode material with PPy/PEG conductive layer, J. Solid State Electr. 14 (2010) 2173-2178. https://doi.org/10.1007/s10008-009-0967-2

[77] H.C. Dinh, S.i. Mho, I.H. Yeo, Electrochemical Analysis of Conductive Polymer-Coated LiFePO$_4$ Nanocrystalline Cathodes with Controlled Morphology, Electroanal. 23 (2011) 2079-2086. https://doi.org/10.1002/elan.201100222

[78] D. Lepage, C. Michot, G.X. Liang, M. Gauthier, S.B. Schougaard, A Soft Chemistry Approach to Coating of LiFePO$_4$ with a Conducting Polymer, Angew. Chem. Int. Ed. 50 (2011) 6884-6887. https://doi.org/10.1002/anie.201101661

[79] P. Zhang, L. Zhang, X. Ren, Q. Yuan, J. Liu, Q. Zhang, Preparation and electrochemical properties of LiNi$_{1/3}$Co$_{1/3}$Mn$_{1/3}$O$_2$–PPy composites cathode materials for lithium-ion battery, Synth. Met. 161 (2011) 1092-1097. https://doi.org/10.1016/j.synthmet.2011.03.021

[80] R. Orinakova, A. Fedorkova, A. Orinak, Effect of PPy/PEG conducting polymer film on electrochemical performance of LiFePO$_4$ cathode material for Li-ion batteries, Chem. Pap. 67 (2013) 860-875. https://doi.org/10.2478/s11696-013-0350-8

[81] D. Cintora-Juarez, C. Perez-Vicente, S. Ahmad, J.L. Tirado, Improving the cycling performance of LiFePO$_4$ cathode material by poly(3,4-ethylenedioxythiopene) coating, RSC Adv. 4 (2014) 26108-26114. https://doi.org/10.1039/C4RA05286B

[82] J.M. Kim, H.S. Park, J.H. Park, T.H. Kim, H.K. Song, S.Y. Lee, Conducting polymer-skinned electroactive materials of lithium-ion batteries: ready for monocomponent electrodes without additional binders and conductive agents, ACS Appl. Mater. Interfaces 6 (2014) 12789-12797. https://doi.org/10.1021/am502736m

[83] H.Y. Yan, W.X. Chen, X.M. Wu, Y.F. Li, Conducting polyaniline-wrapped lithium vanadium phosphate nanocomposite as high-rate and cycling stability cathode for lithium-ion batteries, Electrochim. Acta 146 (2014) 295-300. https://doi.org/10.1016/j.electacta.2014.09.040

[84] J. Cao, G. Hu, Z. Peng, K. Du, Y. Cao, Polypyrrole-coated LiCoO$_2$ nanocomposite with enhanced electrochemical properties at high voltage for lithium-ion batteries, J. Power Sources 281 (2015) 49-55. https://doi.org/10.1016/j.jpowsour.2015.01.174

[85] J. Lee, W. Choi, Surface modification of over-lithiated layered oxides with PEDOT:PSS conducting polymer in lithium-ion batteries, J. Electrochem. Soc. 162 (2015) A743-A748. https://doi.org/10.1149/2.0801504jes

[86] H.Y. Yan, X.M. Wu, Y.F. Li, Preparation and characterization of conducting polyaniline-coated LiVPO$_4$F nanocrystals with core-shell structure and its application in lithium-ion batteries, Electrochim. Acta 182 (2015) 437-444. https://doi.org/10.1016/j.electacta.2015.09.141

[87] F. Wu, J. Liu, L. Li, X. Zhang, R. Luo, Y. Ye, R. Chen, Surface modification of li-rich cathode materials for lithium-ion batteries with a PEDOT:PSS conducting polymer, ACS Appl. Mater. Interfaces 8 (2016) 23095-23104. https://doi.org/10.1021/acsami.6b07431

[88] H. Yan, G. Zhang, Y. Li, Synthesis and characterization of advanced Li$_3$V$_2$(PO$_4$)$_3$ nanocrystals@conducting polymer PEDOT for high energy lithium-ion batteries, App. Surf. Sci. 393 (2017) 30-36. https://doi.org/10.1016/j.apsusc.2016.09.156

[89] Q. Gan, N. Qin, Y. Zhu, Z. Huang, F. Zhang, S. Gu, J. Xie, K. Zhang, L. Lu, Z. Lu, Polyvinylpyrrolidone-induced uniform surface-conductive polymer coating endows Ni-Rich LiNi$_{0.8}$Co$_{0.1}$Mn$_{0.1}$O$_2$ with enhanced cyclability for lithium-ion batteries, ACS Appl. Mater. Interfaces 11 (2019) 12594-12604. https://doi.org/10.1021/acsami.9b04050

[90] N.D. Trinh, M. Saulnier, D. Lepage, S.B. Schougaard, Conductive polymer film supporting LiFePO$_4$ as composite cathode for lithium ion batteries, J. Power Sources 221 (2013) 284-289. https://doi.org/10.1016/j.jpowsour.2012.08.006

Materials Research Forum LLC
https://doi.org/10.21741/9781644900918-2

[91] Y.J. Zhao, Z. Lv, Y. Wang, T. Xu, Combination of Fe-Mn based Li-rich cathode materials and conducting-polymer polypyrrole nanowires with high rate capability, Ionics 24 (2018) 51-60. https://doi.org/10.1007/s11581-017-2166-y

[92] H. Kim, J. Hong, K.Y. Park, H. Kim, S.W. Kim, K. Kang, Aqueous rechargeable Li and Na ion batteries, Chem. Rev. 114 (2014) 11788-11827. https://doi.org/10.1021/cr500232y

[93] H.P. Wong, Synthesis and characterization of polypyrrole/vanadium pentoxide nanocomposite aerogels, J. Mater. Chem. 8 (1998) 1019-1027. https://doi.org/10.1039/a706614g

[94] Z. Tong, H. Yang, L. Na, H. Qu, X. Zhang, J. Zhao, Y. Li, Versatile displays based on a 3-dimensionally ordered macroporous vanadium oxide film for advanced electrochromic devices, J. Mater. Chem. C 3 (2015) 3159-3166. https://doi.org/10.1039/C5TC00029G

[95] H. Zhu, L. Gao, M. Li, H. Yin, D. Wang, Fabrication of free-standing conductive polymer films through dynamic three-phase interline electropolymerization, Electrochem. Commun. 13 (2011) 1479-1483. https://doi.org/10.1016/j.elecom.2011.09.029

[96] K.S. Park, S.B. Schougaard, J.B. Goodenough, Conducting-polymer/iron-redox-couple composite cathodes for lithium secondary batteries, Adv. Mater. 19 (2007) 848-851. https://doi.org/10.1002/adma.200600369

[97] Y.Z. Su, W. Dong, H.H. Zhang, J.H. Song, Y.H. Zhang, K.C. Gong, Poly bis(2-aminophenyloxy)disulfide : A polyaniline derivative containing disulfide bonds as a cathode material for lithium battery, Polymer 48 (2007) 165-173. https://doi.org/10.1016/j.polymer.2006.10.044

[98] G.G. Rodriguez-Calero, M.A. Lowe, S.E. Burkhardt, H.D. Abruna, Electrocatalysis of 2,5-dimercapto-1,3,5-thiadiazole by 3,4-ethylenedioxy-substituted conducting polymers, Langmuir 27 (2011) 13904-13909. https://doi.org/10.1021/la202706s

[99] M. Zhou, J.F. Qian, X.P. Ai, H.X. Yang, Redox-Active Fe(CN)$_6^{4-}$-doped conducting polymers with greatly enhanced capacity as cathode materials for Li-Ion batteries, Adv. Mater. 23 (2011) 4913-4917. https://doi.org/10.1002/adma.201102867

[100] P. Sharma, D. Damien, K. Nagarajan, M.M. Shaijumon, M. Hariharan, Perylene-polyimide-based organic electrode materials for rechargeable lithium batteries, J. Phys. Chem. Lett. 4 (2013) 3192-3197. https://doi.org/10.1021/jz4017359

[101] J. Kim, H.S. Park, T.H. Kim, S.Y. Kim, H.K. Song, An inter-tangled network of redox-active and conducting polymers as a cathode for ultrafast rechargeable

batteries, Phys. Chem. Chem. Phys. 16 (2014) 5295-5300.
https://doi.org/10.1039/c3cp54624a

[102] S.N. Eliseeva, E.V. Alekseeva, A.A. Vereshchagin, A.I. Volkov, P.S. Vlasov, A.S. Konev, O.V. Levin, Nickel-salen type polymers as cathode materials for rechargeable lithium batteries, Macromol. Chem. Phys. 218 (2017) 1700361. https://doi.org/10.1002/macp.201700361

[103] C. O'Meara, M.P. Karushev, I.A. Polozhentceva, S. Dharmasena, H.N. Cho, B.J. Yurkovich, S. Kogan, J.H. Kim, Nickel-salen-type polymer as conducting agent and binder for carbon-free cathodes in lithium-ion batteries, ACS Appl. Mater. Interfaces 11 (2019) 525-533. https://doi.org/10.1021/acsami.8b13742

[104] Y. Kiya, J.C. Henderson, H.D. Abruña, 4-Amino-4H-1, 2, 4-triazole-3, 5-dithiol a modifiable organosulfur compound as a high-energy cathode for lithium-ion rechargeable batteries, J. Electrochem. Soc. 154 (2007) A844-A848. https://doi.org/10.1149/1.2752025

[105] S.R. Deng, L.B. Kong, G.Q. Hu, T. Wu, D. Li, Y.-H. Zhou, Z.-Y. Li, Benzene-based polyorganodisulfide cathode materials for secondary lithium batteries, Electrochim. acta 51 (2006) 2589-2593. https://doi.org/10.1016/j.electacta.2005.07.045

[106] H. Uemachi, Y. Iwasa, T. Mitani, Poly (1, 4-phenylene-1, 2, 4-dithiazol-3, 5'-yl): the new redox system for lithium secondary batteries, Electrochim. acta 46 (2001) 2305-2312. https://doi.org/10.1016/S0013-4686(01)00436-4

[107] G. Yan, J. Li, Y. Zhang, F. Gao, F. Kang, Electrochemical polymerization and energy storage for poly [Ni (salen)] as supercapacitor electrode material, J. Phys. Chem. C 118 (2014) 9911-9917. https://doi.org/10.1021/jp500249t

[108] E.V. Alekseeva, I.A. Chepurnaya, V.V. Malev, A.M. Timonov, O.V. Levin, Polymeric nickel complexes with salen-type ligands for modification of supercapacitor electrodes: impedance studies of charge transfer and storage properties, Electrochim. Acta 225 (2017) 378-391. https://doi.org/10.1016/j.electacta.2016.12.135

[109] E. Dmitrieva, M. Rosenkranz, J.S. Danilova, E.A. Smirnova, M.P. Karushev, I.A. Chepurnaya, A.M. Timonov, Radical formation in polymeric nickel complexes with N2O2 Schiff base ligands: An in situ ESR and UV–vis–NIR spectroelectrochemical study, Electrochim. Acta 283 (2018) 1742-1752. https://doi.org/10.1016/j.electacta.2018.07.131

[110] P.R. Das, L. Komsiyska, O. Osters, G. Wittstock, PEDOT: PSS as a Functional Binder for Cathodes in Lithium Ion Batteries, J. Electrochem. Soc. 162 (2015) A674-A678. https://doi.org/10.1149/2.0581504jes

Materials Research Forum LLC
https://doi.org/10.21741/9781644900918-2

[111] S.N. Eliseeva, O.V. Levin, E.G. Tolstopyatova, E.V. Alekseeva, V.V. Kondratiev, Effect of addition of a conducting polymer on the properties of the LiFePO4-based cathode material for lithium-ion batteries, Russ. J. App. Chem. 88 (2015) 1146-1149. https://doi.org/10.1134/S1070427215070071

[112] H. Zhong, A. He, J. Lu, M. Sun, J. He, L. Zhang, Carboxymethyl chitosan/conducting polymer as water-soluble composite binder for LiFePO$_4$ cathode in lithium ion batteries, J. Power Sources 336 (2016) 107-114. https://doi.org/10.1016/j.jpowsour.2016.10.041

[113] X. Ma, S. Zou, A. Tang, L. Chen, Z. Deng, B.G. Pollet, S. Ji, Three-dimensional hierarchical walnut kernel shape conducting polymer as water soluble binder for lithium-ion battery, Electrochimica Acta 269 (2018) 571-579. https://doi.org/10.1016/j.electacta.2018.03.031

[114] K.A. Vorobeva, S.N. Eliseeva, R.V. Apraksin, M.A. Kamenskii, E.G. Tolstopjatova, V.V. Kondratiev, Improved electrochemical properties of cathode material LiMn$_2$O$_4$ with conducting polymer binder, J. Alloy Compd. 766 (2018) 33-44. https://doi.org/10.1016/j.jallcom.2018.06.324

[115] L. Shacklette, J. Toth, N. Murthy, R. Baughman, Polyacetylene and polyphenylene as anode materials for nonaqueous secondary batteries, J. Electrochem. Soc. 132 (1985) 1529-1535. https://doi.org/10.1149/1.2114159

[116] B. Coffey, P.V. Madsen, T.O. Poehler, P.C. Searson, High charge-density conducting polymer graphite fiber-composite electrodes for battery applications, J. Electrochem. Soc. 142 (1995) 321-325. https://doi.org/10.1149/1.2043991

[117] M. Endo, C. Kim, T. Hiraoka, T. Karaki, M. Matthews, S. Brown, M. Dresselhaus, Li storage behavior in polyparaphenylene (PPP)-based disordered carbon as a negative electrode for Li ion batteries, Molecular Crystals and Liquid Crystals Science and Technology. Section A. Mol. Cryst. Liq. Cryst. 310 (1998) 353-358. https://doi.org/10.1080/10587259808045361

[118] X.M. He, J.G. Ren, L. Wang, W.H. Pu, C.Y. Jiang, C.R. Wan, Synthesis of PAN/SnCl$_2$ composite as Li-ion battery anode material, Ionics 12 (2006) 323-326. https://doi.org/10.1007/s11581-006-0051-1

[119] J. Chen, Y. Liu, A.I. Minett, C. Lynam, J. Wang, G.G. Wallace, Flexible, aligned carbon nanotube/conducting polymer electrodes for a lithium-ion battery, Chem. Mater. 19 (2007) 3595-3597. https://doi.org/10.1021/cm070991g

[120] C. Reynaud, C. Boiziau, C. Juret, S. Leroy, J. Perreau, G. Lecayon, Valence electronic structure of a thin film of polyacrylonitrile and its pyrolyzed derivatives, Synth. Met. 11 (1985) 159-165. https://doi.org/10.1016/0379-6779(85)90061-X

[121] W. Lee, E. Choi, A. Ovchinnikov, Y. Park, K. Liou, D. Kim, K. Chung, Electrical transport of the pyrolyzed materials; polyacenic and polyacrylonitrile, Synth. Met. 57 (1993) 5075-5080. https://doi.org/10.1016/0379-6779(93)90865-T

[122] S. Ng, J. Wang, Z. Guo, J. Chen, G. Wang, H. Liu, Single wall carbon nanotube paper as anode for lithium-ion battery, Electrochim. Acta 51 (2005) 23-28. https://doi.org/10.1016/j.electacta.2005.04.045

[123] Y. Lu, L. Yu, X.W.D. Lou, Nanostructured conversion-type anode materials for advanced lithium-ion batteries, Chem 4 (2018) 972-996. https://doi.org/10.1016/j.chempr.2018.01.003

[124] C.K. Chan, H. Peng, G. Liu, K. McIlwrath, X.F. Zhang, R.A. Huggins, Y. Cui, High-performance lithium battery anodes using silicon nanowires, Nat. Nanotechnol. 3 (2008) 31. https://doi.org/10.1038/nnano.2007.411

[125] Y. Xiao, X. Wang, Y. Xia, Y. Yao, E. Metwalli, Q. Zhang, R. Liu, B. Qiu, M. Rasool, Z. Liu, J.Q. Meng, L.D. Sun, C.H. Yan, P. Muller-Buschbaum, Y.J. Cheng, Green facile scalable synthesis of titania/carbon nanocomposites: New use of old dental resins, ACS Appl. Mater. Interfaces 6 (2014) 18461-18468. https://doi.org/10.1021/am506114p

[126] X. Wang, J.Q. Meng, M. Wang, Y. Xiao, R. Liu, Y. Xia, Y. Yao, E. Metwalli, Q. Zhang, B. Qiu, Z. Liu, J. Pan, L.D. Sun, C.H. Yan, P. Muller-Buschbaum, Y.J. Cheng, Facile scalable synthesis of TiO_2/carbon nanohybrids with ultrasmall TiO_2 nanoparticles homogeneously embedded in carbon matrix, ACS Appl. Mater. Interfaces 7 (2015) 24247-24255. https://doi.org/10.1021/acsami.5b07784

[127] P.M. Dziewonski, M. Grzeszczuk, Towards TiO_2-conducting polymer hybrid materials for lithium ion batteries, Electrochim. Acta 55 (2010) 3336-3347. https://doi.org/10.1016/j.electacta.2010.01.043

[128] X. Shen, D. Mu, S. Chen, B. Wu, F. Wu, Enhanced electrochemical performance of ZnO-loaded/porous carbon composite as anode materials for lithium ion batteries, ACS Appl. Mater. Interfaces 5 (2013) 3118-3125. https://doi.org/10.1021/am400020n

[129] N. Li, S. Jin, Q. Liao, C. Wang, ZnO anchored on vertically aligned graphene: binder-free anode materials for lithium-ion batteries, ACS Appl. Mater. Interfaces 6 (2014) 20590-20596. https://doi.org/10.1021/am507046k

[130] G. Zhang, S. Hou, H. Zhang, W. Zeng, F. Yan, C.C. Li, H. Duan, High-performance and ultra-stable lithium-ion batteries based on MOF-derived ZnO@ZnO quantum Dots/C core-shell nanorod arrays on a carbon cloth anode, Adv. Mater. 27 (2015) 2400-2405. https://doi.org/10.1002/adma.201405222

[131] G.L. Xu, Y. Li, T. Ma, Y. Ren, H.-H. Wang, L. Wang, J. Wen, D. Miller, K. Amine, Z. Chen, PEDOT-PSS coated ZnO/C hierarchical porous nanorods as ultralong-life anode material for lithium ion batteries, Nano Energy 18 (2015) 253-264. https://doi.org/10.1016/j.nanoen.2015.10.020

[132] L. Zhao, Y.S. Hu, H. Li, Z. Wang, L. Chen, Porous $Li_4Ti_5O_{12}$ coated with n-doped carbon from ionic liquids for Li-ion batteries, Adv. Mater. 23 (2011) 1385-1388. https://doi.org/10.1002/adma.201003294

[133] G.N. Zhu, Y.G. Wang, Y.Y. Xia, Ti-based compounds as anode materials for Li-ion batteries, Energy Environ. Sci. 5 (2012) 6652-6667. https://doi.org/10.1039/c2ee03410g

[134] L. Zheng, X. Wang, Y. Xia, S. Xia, E. Metwalli, B. Qiu, Q. Ji, S. Yin, S. Xie, K. Fang, Scalable in situ synthesis of $Li_4Ti_5O_{12}$/carbon nanohybrid with supersmall Li4Ti5O12 nanoparticles homogeneously embedded in carbon matrix, ACS Appl. Mater. Interfaces 10 (2018) 2591-2602. https://doi.org/10.1021/acsami.7b16578

[135] H. Zhang, Q. Deng, C. Mou, Z. Huang, Y. Wang, A. Zhou, J. Li, Surface structure and high-rate performance of spinel $Li_4Ti_5O_{12}$ coated with N-doped carbon as anode material for lithium-ion batteries, J. Power Sources 239 (2013) 538-545. https://doi.org/10.1016/j.jpowsour.2013.03.013

[136] D. Xu, P.F. Wang, R. Yang, Conducting polythiophene-wrapped $Li_4Ti_5O_{12}$ spinel anode material for ultralong cycle-life Li-ion batteries, Ceram. Int. 43 (2017) 4712-4715. https://doi.org/10.1016/j.ceramint.2016.12.116

[137] Z.J. Du, S.C. Zhang, T. Jiang, X.M. Wu, L. Zhang, H. Fang, Facile synthesis of SnO_2 nanocrystals coated conducting polymer nanowires for enhanced lithium storage, J. Power Sources 219 (2012) 199-203. https://doi.org/10.1016/j.jpowsour.2012.07.052

[138] Y. Yao, N. Liu, M.T. McDowell, M. Pasta, Y. Cui, Improving the cycling stability of silicon nanowire anodes with conducting polymer coatings, Energy Environ. Sci. 5 (2012) 7927-7930. https://doi.org/10.1039/c2ee21437g

[139] H. Wu, G.H. Yu, L.J. Pan, N.A. Liu, M.T. McDowell, Z.A. Bao, Y. Cui, Stable Li-ion battery anodes by in-situ polymerization of conducting hydrogel to conformally coat silicon nanoparticles, Nat. Commun. 4 (2013) 1943. https://doi.org/10.1038/ncomms2941

[140] S. Xun, X. Song, V. Battaglia, G. Liu, Conductive polymer binder-enabled cycling of pure tin nanoparticle composite anode electrodes for a lithium-ion battery, J. Electrochem. Soc. 160 (2013) A849-A855. https://doi.org/10.1149/2.087306jes

[141] D. Shao, H. Zhong, L. Zhang, Water-Soluble Conductive Composite Binder Containing PEDOT:PSS as Conduction Promoting Agent for Si Anode of Lithium-Ion Batteries, ChemElectroChem 1 (2014) 1679-1687. https://doi.org/10.1002/celc.201402210

[142] T.M. Higgins, S.H. Park, P.J. King, C. Zhang, N. MoEvoy, N.C. Berner, D. Daly, A. Shmeliov, U. Khan, G. Duesberg, V. Nicolosi, J.N. Coleman, A commercial conducting polymer as both binder and conductive additive for silicon nanoparticle-based lithium-ion battery negative electrodes, ACS Nano 10 (2016) 3702-3713. https://doi.org/10.1021/acsnano.6b00218

[143] E. Park, J. Kim, D.J. Chung, M.-S. Park, H. Kim, J.H. Kim, Si/SiO$_x$-conductive polymer core-shell nanospheres with an improved conducting path preservation for lithium-ion battery, ChemSusChem 9 (2016) 2754-2758. https://doi.org/10.1002/cssc.201600798

[144] H. Zhao, A. Du, M. Ling, V. Battaglia, G. Liu, Conductive polymer binder for nano-silicon/graphite composite electrode in lithium-ion batteries towards a practical application, Electrochim. Acta 209 (2016) 159-162. https://doi.org/10.1016/j.electacta.2016.05.061

[145] H. Zhao, Y. Fu, M. Ling, Z. Jia, X. Song, Z. Chen, J. Lu, K. Amine, G. Liu, Conductive polymer binder-enabled SiO-Sn$_x$Co$_y$C$_z$ anode for high-energy lithium-ion batteries, ACS Appl. Mater. Interfaces 8 (2016) 13373-13377. https://doi.org/10.1021/acsami.6b00312

[146] H. Chu, K. Lee, S. Lim, T.H. Kim, Enhancing the performance of a silicon anode by using a new conjugated polymer binder prepared by direct arylation, Macromol. Res. 26 (2018) 738-743. https://doi.org/10.1007/s13233-018-6106-0

[147] B. Liu, P. Soares, C. Checkles, Y. Zhao, G. Yu, Three-dimensional hierarchical ternary nanostructures for high-performance Li-ion battery anodes, Nano Lett 13 (2013) 3414-3419. https://doi.org/10.1021/nl401880v

[148] Z. Chen, J.W.F. To, C. Wang, Z.D. Lu, N. Liu, A. Chortos, L.J. Pan, F. Wei, Y. Cui, Z.N. Bao, A three-dimensionally interconnected carbon nanotube-conducting polymer hydrogel network for high-performance flexible battery Electrodes, Adv. Energy Mater. 4 (2014). https://doi.org/10.1002/aenm.201400207

[149] W.F. Ren, J.T. Li, Z.G. Huang, L. Deng, Y. Zhou, L. Huang, S.G. Sun, Fabrication of Si nanoparticles@conductive carbon framework@polymer composite as high-areal-capacity anode of lithium-ion batteries, ChemElectroChem 5 (2018) 3258-3265. https://doi.org/10.1002/celc.201800834

[150] G. Liu, S. Xun, N. Vukmirovic, X. Song, P. Olaldeosite as high-areal-capacity anode of lithium-ion battlexible berfh tailored electronic structure for high capacity

lithium battery electrodes, Adv. Mater. 23 (2011) 4679-4683.
https://doi.org/10.1002/adma.201102421

[151] I. Kovalenko, B. Zdyrko, A. Magasinski, B. Hertzberg, Z. Milicev, R. Burtovyy, I.
Luzinov, G. Yushin, A major constituent of brown algae for use in high-capacity Li-
ion batteries, Science 334 (2011) 75-79. https://doi.org/10.1126/science.1209150

[152] J. Wu, X. Rui, G. Long, W. Chen, Q. Yan, Q. Zhang, Pushing up lithium storage
through nanostructured polyazaacene analogues as anode, Angew. Chem. Int. Ed. 54
(2015) 7354-7358. https://doi.org/10.1002/anie.201503072

[153] J. Wu, X. Rui, C. Wang, W.B. Pei, R. Lau, Q. Yan, Q. Zhang, Nanostructured
conjugated ladder polymers for stable and fast lithium storage anodes with high-
capacity, Adv. Energy Mater. 5 (2015) 1402189.
https://doi.org/10.1002/aenm.201402189

[154] J. Xie, X. Rui, P. Gu, J. Wu, Z.J. Xu, Q. Yan, Q. Zhang, Novel conjugated ladder-
structured oligomer anode with high lithium storage and long cycling capability, ACS
Appl. Mater. Interfaces 8 (2016) 16932-16938.
https://doi.org/10.1021/acsami.6b04277

[155] J. Hu, F.F. Jia, Y.F. Song, Engineering high-performance
polyoxometalate/PANI/MWNTs nanocomposite anode materials for lithium ion
batteries, Chem. Eng. J. 326 (2017) 273-280.
https://doi.org/10.1016/j.cej.2017.05.153

[156] Q.S. Liao, H.Y. Hou, J.X. Duan, S. Liu, Y. Yao, Z.P. Dai, C.Y. Yu, D.D. Li,
Composite sodium rho-toluene sulfonate-polypyrrole-iron anode for a lithium-ion
battery, J. App. Polym. Sci. 134 (2017) 44935. https://doi.org/10.1002/app.44935

[157] G. Sandu, B. Ernould, J. Rolland, N. Cheminet, J. Brassinne, P.R. Das, Y.
Filinchuk, L.H. Cheng, L. Komsiyska, P. Dubois, S. Melinte, J.F. Gohy, R. Lazzaroni,
A. Vlad, Mechanochemical synthesis of PEDOT:PSS hydrogels for aqueous
formulation of Li-ion battery electrodes, ACS Appl. Mater. Interfaces 9 (2017)
34865-34874. https://doi.org/10.1021/acsami.7b08937

[158] A.D. Schlüter, Ladder polymers: The new generation, Adv. Mater. 3 (1991) 282-
291. https://doi.org/10.1002/adma.19910030603

[159] F. Arnold, R. Van Deusen, Unusual film-forming properties of aromatic
heterocyclic ladder polymers, J. App. Polym. Sci. 15 (1971) 2035-2047.
https://doi.org/10.1002/app.1971.070150820

[160] P. Bornoz, M.S. Prévot, X. Yu, N.s. Guijarro, K. Sivula, Direct light-driven water
oxidation by a ladder-type conjugated polymer photoanode, J. Am. Chem. Soc. 137
(2015) 15338-15341. https://doi.org/10.1021/jacs.5b05724

Lithium-ion Batteries - Materials and Applications   Materials Research Forum LLC
Materials Research Foundations **80** (2020) 28-62   https://doi.org/10.21741/9781644900918-2

[161] Z.X. Cai, M.A. Awais, N. Zhang, L.P. Yu, Exploration of Syntheses and Functions of Higher Ladder-type pi-Conjugated Heteroacenes, Chem. 4 (2018) 2538-2570. https://doi.org/10.1016/j.chempr.2018.08.017

[162] M.H. Hoang, G.E. Park, D.L. Phan, T.T. Ngo, T.V. Nguyen, C.G. Park, M.J. Cho, D.H. Choi, Synthesis of conjugated wide-bandgap copolymers bearing ladder-type donating units and their application to non-fullerene polymer solar cells, Macromol. Res. 26 (2018) 844-850. https://doi.org/10.1007/s13233-018-6128-7

[163] Y.L. Yin, S.Y. Zhang, D.Y. Chen, F.Y. Guo, G. Yu, L.C. Zhao, Y. Zhang, Synthesis of an indacenodithiophene-based fully conjugated ladder polymer and its optical and electronic properties, Polym. Chem. 9 (2018) 2227-2231. https://doi.org/10.1039/C8PY00351C

[164] J.R. Cheng, B. Li, X.J. Ren, F. Liu, H.C. Zhao, H.J. Wang, Y.G. Wu, W.P. Chen, X.W. Ba, Highly twisted ladder-type backbone bearing perylene diimides for non-fullerene acceptors in organic solar cells, Dyes Pigments 161 (2019) 221-226. https://doi.org/10.1016/j.dyepig.2018.09.042

[165] F.E. Arnold, R. Van Deusen, Preparation and properties of high molecular weight, soluble oxobenz [de] imidazobenzimidazoisoquinoline ladder polymer, Macromolecules 2 (1969) 497-502. https://doi.org/10.1021/ma60011a009

[166] F. Dawans, C. Marvel, Polymers from ortho aromatic tetraamines and aromatic dianhydrides, J. Polym. Sci. Pol. Chem. 3 (1965) 3549-3571. https://doi.org/10.1002/pol.1965.100031019

Chapter 3

2D Transition Metal Dichalcogenides for Lithium-ion Batteries

S.V. Prabhakar Vattikuti

School of Mechanical Engineering, Yeungnam University, Gyeongsan, 712749, South Korea

drprabu@ynu.ac.kr

Abstract

Transition metal dichalcogenides (TMDs) materials offer exciting prospective applications in energy storage particularly as electrode resources for lithium-ion batteries (LIBs), which lately received enormous consideration owing to their unique 2D layered features appropriate for effective charge storage and conversion. However, several important and practical challenges are still need to be resolved for LIBs to become suitable for commercial use. This chapter presents a timely review, detailing the perspectives on the latest advancements and future directions in using TMDs for LIBs research. The chapter is divided into three parts: (i) TMDs materials as electrode; (ii) TMDs based hybrids; and (iii) viewpoint and perspectives.

Keywords

Lithium-Ion Batteries, Layered Materials, Transition Metal Dichalcogenides, Molybdenum Disulfide, Tungsten Disulfide

Contents

1.   Introduction

The rapidly increasing global demand for sustainable energy as well as the conventional environmental impacts of energy consumption necessitates the need for safer and more cost-effective energy storage technologies. Therefore, there is huge plea for safe, clean and green sustainable energy sources. To address this issue, a huge attention has been actively focused in the arena of energy storage as well as conversion [1]. Well knowingly, the lithium-ion batteries (LIBs) have been comprehensively exploited in various fields of applications and are of favored choice at this present scenario. Over the years, lots of improvements have been taken to efficiently charged and discharged. For instance, modern hybrid electric vehicle (MHEV) demands the deployment of high capacity LIBs, however high energy and power density parameters are difficult to satisfy and thus require more high energy batteries. The LIBs suffer from low powder density and cyclic stability; hence attempts were made to improve LIBs to contain high energy as well as to meet the desirable powder density with remarkable cyclic stability.

To date, extensive research efforts have been focused on enhancing the energy density of LIBs by using various nanomaterials, including transition metal oxides (TMOs) or dichalcogenide (TMDs) and metal oxides. Efficient electrodes should possess good electronic conductivity and be adept in accommodating Li ions in a fast rate. In the periodic table, group IV-VII are dominantly layered in nature, while, about of group VIII-X TMDs are mostly originating in other morphological materials and have been viewed to be possibly used for energy storage-conversion. Other morphological TMD such as pyrite (FeS_2) has appeared superior in the energy storage. This non-layered TMD FeS_2 is well recognized as a substitute for gold and confirms a cubic configuration of which the unit cell is comprises of a face-centered Fe coupled with S ions. The Fe ions are enclosed with adjacent six sulfur neighbors in a partial octahedral plan.

Transition metal dichalcogenides (TMDs) commonly known as MX_2 (M-transition metal i.e. Mo, W and X-chalcogen i.e. S, Se) have currently received remarkable interest attributing to their exclusive and attractive electronic, optical and mechanical characteristics [2], their molecular structures are shown in Fig.1. In this type of complexes, the atoms within the layers are assembled via robust covalent bonds (X-M-X) and the distinct sheets are attached by van der Waals connections creating a layered by layered assembly. These features are advantageous for the diffusion of Li ions making

TMDs a plausible anode candidate material. For instance, the TMD WSe_2 has a covered sandwich structure comprising of Se-W-Se with a far greater interlayer spacing (0.65 nm) compared to graphite (0.33 nm), allowing the efficient diffusion of Li ions. Novel MX_2 materials possess advanced inherent electronic conductivity, greater interlayer spacing and small potential plateau compared to commercial ones, thus hold more benefits for emerging secondary batteries. The following equations exemplify the mechanism of Li/Na–ion storage [2].

$$MX_2 + 4A^+ + 4e^- \leftrightarrow M + 2A_2X \text{ (A i.e. Li, and Na)} \tag{1}$$

Equation (1) is comprehensive with the subsequent double-step electrochemical responses,

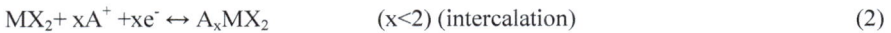

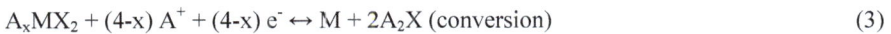

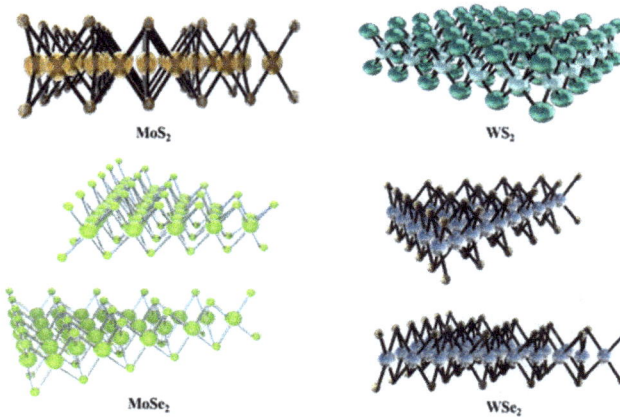

$$MX_2 + xA^+ + xe^- \leftrightarrow A_xMX_2 \qquad (x<2) \text{ (intercalation)} \tag{2}$$

$$A_xMX_2 + (4-x) A^+ + (4-x) e^- \leftrightarrow M + 2A_2X \text{ (conversion)} \tag{3}$$

Figure 1. Popular Layered Transition metal dichalcogenides (TMDs) materials for LIBs.

A potential electrode should retain good electric conductivity and be accomplished of accepting Li ions in a fast rate. To attain this, an efficient route is to shorten the ion diffusion distance within a conductive matrix-like carbonaceous materials such as active carbon and graphene. But this approach comes at the cost of electrodes tape density. Importantly, searching materials which hold inherent high electronic conductivity and superficial ion diffusion pathways are of crucial importance in achieving superior lithium storage performance. Although outstanding power handling and decent gravimetric architectures including graphene aerogels, carbon nanotube aerogels, sponges, and foams are more highlighted in the practical situation, they are far from expectation and still need complete understanding of the electrochemical reactions.

The mechanism of MX_2 lithiation/delithiation has trigged a notable consideration in the energy storage community. Investigators strongly agreed that on the primary discharge process, the MX_2 is reconverted into metallic M nanoparticles after interaction with the Li ions in the Li_2X matrix [3, 4]. However, no consistency has been established regarding the fundamental knowledge underlying the chemistry of the subsequent processes. Contrasting arguments have been documented concerning the conversion rate of MX_2 into Li_2X and M. It has been reported that this process is irreversible, a reaction ascribed to Li-X batteries. While other report showed that the discharge yields of Li_2X and M are converted into MX_2, suggesting a reversible conversion mechanism. With these contrasting reports, further investigation exploring the mechanism of MX_2 lithium storage must be carried out to clarify this process.

2. MoS_2-based anode materials for LIBs

MoS_2 is a one of the typical layered TMDs with a maximum theoretic capacity of 670 $mAhg^{-1}$ for LIBs, a value greater than the standard graphite (372 $mAhg^{-1}$). The higher capacity value of MoS_2 is attributable to the relocation of its four electrons during charge/discharge process. However, the use of MoS_2 for LIBs electrode materials is bounded by several intrinsic demerits. For example, the low electronic conductivity, a factor attributed to its weak van der Waals interaction. This weak interaction may reduce the transport of electrons and ions particularly between layers in a c-direction, resulting to a capacity loss during continuous use. Among the TMDs, MoS_2 is one of the well-known van der Waals solids, which possesses distinctive physical characteristics in single-to-few layer forms. MoS_2-based materials have been widely employed as electrode materials for LIBs and lithium-sulfur batteries (LSBs) [5]. The MoS_2 lattice spacing can deliver a remarkable space for accommodating Li-ions and has a hole mobility of 741 $cm^2V^{-1}S^{-1}$, making MoS_2 favorable to be employed as energy materials for LIBs. Some reports revealed that MoS_2 has been used as the cathode materials for LIBs [6, 7]. However over the years, the application of MoS_2 has been extended from cathode to anode materials. The applicability of MoS_2 as anode materials for LIBs has drawn enormous attention, attributing to its distinctive electrochemical behavior. For an anode material, a standard Li ion insertion-extraction process into a perfect MoS_2 can be disclosed as follows: [8]

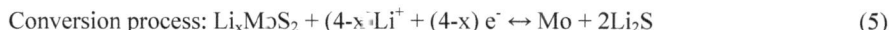

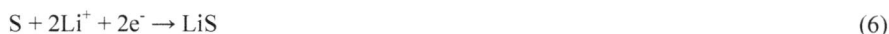

Intercalation process : $MoS_2 + xLi^+ + xe^- \leftrightarrow Li_xMoS_2$ (4)

Conversion process: $Li_xMoS_2 + (4-x)Li^+ + (4-x) e^- \leftrightarrow Mo + 2Li_2S$ (5)

$S + 2Li^+ + 2e^- \rightarrow LiS$ (6)

Two dimensional (2D) layered TMDs exhibited a striking difference with regards to electronic structure as compared to a bulk three dimensional (3D) TMDs. Exfoliated 2D TMDs nanolayers have been viewed as attractive anode materials for LIBs due to the stacked arrangement of nanosheets that could accommodate the tidal volume generated from cyclic reversible volume variations during Li cycling. However, optimization and further work regarding the synthetic procedure of the 2D TMDs nanosheets and rational intended 3D structure are of abundant significance that needs to be addressed. Currently, MoS_2-based nanocrystals with various structural models have been achieved by chemical vapor deposition (CVD), pyrolysis, microwave colloidal synthesis, solvo-hydrothermal reaction by strong reductant, and chemical/electrochemical route [9, 10]. Among these procedures, the CVD in an extraordinary vacuum condition creates it challenging to realize the progress at minimal cost due to the complexity of the synthetic procedure, which involves high temperature and harsh environmental conditions. Furthermore, the poor electrical conductivity of the MoS_2 has been recognized to moderately limit its exploitation for energy storing. Hence, it is required to advance a facile, modest, low-cost and globally friendly process to obtain the TMDs-based nanoconfigurations with admirable electrochemical behavior for LIBs.

MoS_2 employed as anode for LIBs has been observed limited to use because of the polysulfide shuttling effect which can cause fast electrode deteriorate [11]. Up to now, several reports have been documented exploring the possible approaches in orchestrating a well-suited MoS_2 with improved characteristic conductivity such as coupling with conductive carbon or non-carbon materials. Combining with other materials, particularly with different 0D to 3D carbon materials is a famous strategy to boost the electro-chemical conductivity of MoS_2 and to provide supports in hybrid, preserving a constant structure during the long term and high current densities. It is undoubtedly considered as a well-organized strategy to facilitate electro-chemical activity of carbon@MoS_2 hybrids although the electrochemical mechanism between the MoS_2 and carbon is still unclear [12, 13].

The unique architecture of hollow carbon spheres offers an added benefit for electrode-electrolyte interaction allowing the reduction of ion diffusion distances and the defending of volume changes. Wan et al. [14] reported the engagement of a hydrothermally synthesized quasi-hollow microspheres of MoS_2 which are anchored on porous carbon as anode for LIBs. These $C@MoS_2$ microspheres demonstrated high cycling stability up to 95% capacity retention afterward 100 cycles with good rate behavior of 560 mAhg^{-1} at 5 Ag^{-1}. Importantly, coupling MoS_2 with three dimensional (3D) carbon materials like graphene-aerogel paper and carbon sheet-like links is the best choice for the synthesis of MoS_2-based materials for LIBs. For example, a 3D graphene along with few-layered MoS_2 nanosheets demonstrated a remarkable capacity of 1077 mAhg^{-1} at 100 mAg^{-1} at 150 cycles. 3D graphene aerogel possesses a high specific surface area, interconnected electron pathways and lightweight features making it a prominent scaffolding material for electrodes [15]. In another report presented by Zhou et al. [16], in-situ synthesized MoS_2 nanosheets fastened on 3D mesoporous carbon nanosheets frameworks showed superior electrochemical properties and long-time cycling stability of 676 mAhg^{-1} at 2Ag^{-1} for 520 sequences.

Some of the non-carbonaceous materials including, TiO_2, SnO_2, Fe_3O_4, Ag, or Si, can also improve the electrochemical performance of MoS_2. For instance, TiO_2 showed the lowest volume variation (i.e. less than 4%) as compared to Li-ion, has the greatest recycling stability in LIBs amid all transition metal oxides as well as sulfide-type anode materials, and has been demonstrated to enhance their stability particularly at robust current densities when coupled with MoS_2-based materials [17, 18]. This TiO_2/MoS_2 hetereostructure demonstrates a revocable capacity of 710 mAhg^{-1} at 100 mAg^{-1} after 100 progressions and showed good stability of capacity retention. Taken together, it is apparent that the aforementioned approaches like the crafting of MoS_2 constraints, developing special morphologies and the coupling with carbonaceous or metal oxide materials are all capable of improving the properties and quality of MoS_2-typed anode ingredients for LIBs. To date, few studies have proposed that merging two or more of these approaches can obtain even superior electrochemical properties than single approach alone. This claim has been demonstrated by Yu et al. [19] by developing a unique N-doped carbon@MoS_2 box like-nanoarchitecture consisting of graded MoS_2 box-like structures. Adding the N-doped carbon cells conferred an enhanced electric conductivity together with the anchored tiny MoS_2 nanosheets forming all together a nanosheet-on-nanobox structure. As associated with the ordered MoS_2 boxes, N-fixed carbon@MoS_2 boxes facilitate enhanced charge-discharge cycling activities with cyclic stabilities (1000 mAhg^{-1} at 200 cycles).

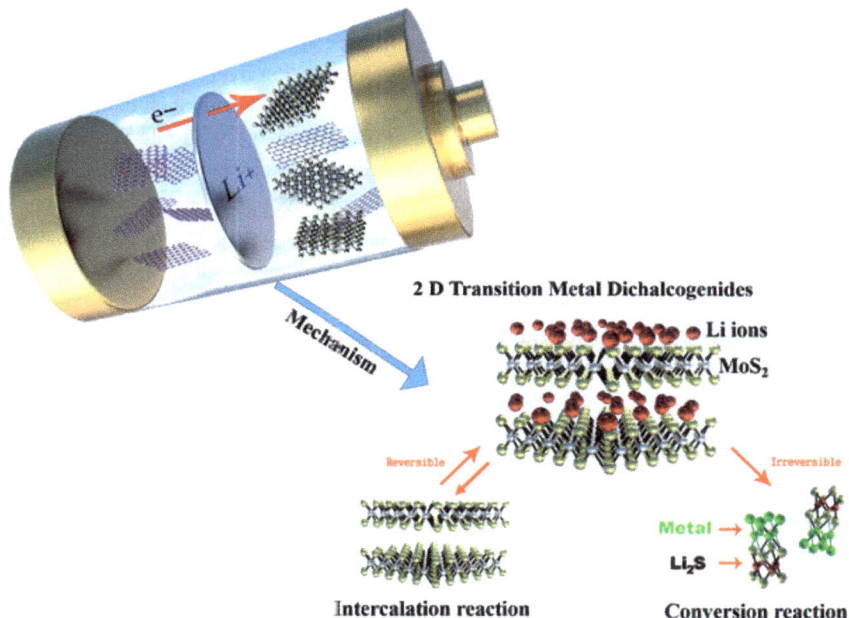

Figure 2. Representation of TMDs based electrode materials for LIBs and their intercalation/conversion reactions.

On the other hand, an advanced C-Si supported MoS_2 ternary construction consists of few-layered C@MoS_2 microspheres anchored on Si nanoparticles was reported by Kang et al. [20]. In this report, Si nanoparticles play a significant part in enlightening the electrochemical behavior owing to their small release potential and enormously extraordinary reversible capacity. The carbon fascinated MoS_2 microspheres serve as the supports, thus maintaining the structural feasibilities during charge-discharge process and also aid in optimizing the benefits of Si nanoparticles. As a result, these distinctive hierarchical structures demonstrate superior capability of 1020 mAhg^{-1} at 1000 mAg^{-1} after 100cycles as compared to bare MoS_2/C microsphere (718 mAhg^{-1}). Similar work has been reported by Shan et al. [21], where 3DMoS_2@carbon fibers exhibited high reversible areal capacity of 5.2 mAh cm^{-2} on 2.5 mA.cm^{-2} with superior rate behavior.

Materials Research Forum LLC
https://doi.org/10.21741/9781644900918-3

The authors prepared carbon fibers via merely carbonizing the waste cotton employed as carbon source hence, reflecting a cost effective and eco-friendly approach. Studies accompanied by Xue et al. [22] and Wang et al. [23] revealed that MoS_2 microsphere/N-doped carbon demonstrated excellent reversible capacity of 715 mAhg^{-1} at 100 mAg^{-1} after 50 sequences with remarkable cycling stability and that hydrothermally synthesized sheet-like MoS_2 developed on the surfaces of carbon and TiO_2 core-shell nanofibers have great reversible specific capacity of 1460 mAhg^{-1} at 100 mAg^{-1} with rate performance of 928 mAhg^{-1} at 2Ag^{-1}. This binder free anode materials are suitable for flexible LIBs.

Recent studies revealed that rational design of effective and durable anode materials are potentially significant for high electrochemical performance of LIBs. In this view point, many researchers have developed conductive materials that are structurally modified such as graphene, through the introduction of noble metals or doping of non-precious metal to improve their electrochemical reactions. For instance, Yang et.al. [24] designed MoS_2/N with S co-doped graphene via simple electrochemical exfoliation combined with hydrothermal reactions of the precursors. A 3D interconnection of few-layered MoS_2 and graphene have been achieved by this synthetic procedure, a structure that relieves the restacking of binary components and thereby accelerated the transport of electrons and enhancing Li-ion storage capacity as an electrode for batteries. This anode exhibited good revocable capability of 1012 mAhg^{-1} after cycling 300 times at 0.5 Ag^{-1}. It has also delivered a maximum energy density of 890 Whkg^{-1} along with the powder compactness of 130 W kg^{-1}.

Wang et al. [25] invented nanosized porous $MoS_{2(1-x)}Se_{2x}$ hybrid encapsulated in graphene sheets to form a mesoporous heterostructure (i.e. meso-$MoS_{2(1-x)}Se_{2x}$/rGO hybrid) consisting of perpendicularly grown 2D sheet-like structure on graphene layers. The meso-$MoS_{2(1-x)}Se_{2x}$/rGO hybrid possessed an improved electrochemical performance for LIBs, ascribed to the benefits of 2D sheet-like alignment and the sulfur-selenium joined effect. Adding Se may confer additional active diffusion path for both Li ions and electrons during the charge/discharge reactions by possibly broadening the positioning of the (002) crystal plane. Moreover, the introduction of Se in MoS_2 may auxiliary induce the transition of meso-$MoS_{2(1-x)}Se_{2x}$/rGO hybrid from 2H phase to further stable metallic 1T phase of MoS_2. The meso-$MoS_{2(1-x)}Se_{2x}$/rGO hybrid exhibited capable rate performances and respectable cycling stability with an best capacity of 830 mAhg^{-1} at a current density of 100 mAg^{-1} for 150th cycle. This hybrid maintained a capacity of 415.3 mA g^{-1} at enormous current density is 1000 mAg^{-1} creating them a prominent candidate for different types of batteries. In a parallel study, Zhu et al. [26] was able to produce MoS_2/N-doped carbon (i.e. MoS_2/N-C) from simplistic instantaneous vapor sulfurization and carbonization of MoO_x/aniline nanorods and has employed it as anode. This MoS_2/N-

C electrode displayed remarkable rate capability of 667 mAhg^{-1} at 10 Ag^{-1} and outstanding cyclic stability of 1110 mAhg^{-1} at 1Ag^{-1}. This outstanding electrochemical performance exhibited by MoS$_2$/N-C electrode was assigned to the synergistic effects of MoS$_2$ with extended interlayer spacing and N-doped carbon. In addition, the ample voids produced by the distended interlayer spacing of MoS$_2$ promote the diffusion of Li-ion and can enhance the volume changes during the charge/discharge responses, creating further active sites for Li ion storage.

3. WS$_2$-based anode materials for LIBs

Among the graphene-like two dimensional (2D) TMDs, tungsten disulfide (WS$_2$) has received growing devotion owing to its good mechanical, electric, and optical characteristics [27]. Layered WS$_2$ is self-possessed of sandwich structure arranged via the weak van der Waals forces and robust covalent links in-plane among S and W atoms. This structural configuration is advantageous for lithium ion storing with a theoretical capability of 433 mAhg^{-1}, allowing intercalation of Li atoms in between the layered structure easily [28]. WS$_2$ has a great density of 7.6g cm^{-3} making its volumetric powder density four folds greater than that of standard graphite [29]. WS$_2$ is encompassed of three arranged atomic sheets (S-W-S) detained together via weak Van der Waals strength similar to MoS$_2$, although the interactions among WS$_2$ sheets are moderately weaker than that of MoS$_2$. In addition, WS$_2$ owns an advanced intrinsic conductivity in comparison to MoS$_2$ [30]. These features can provide lithium ions more access to easily intercalate/deintercalate. However, WS$_2$ also suffers with low electrical conductivity as well as undependable stability during long cycles similar to MoS$_2$, features that limit its application in LIBs.

In order to overcome these limitations, two methods are commonly adapted including the coupling of WS$_2$ with carbon-based materials and phase alteration from semiconducting WS$_2$ to metallic phase. Controlling the crystalline favored orientation including (001) plane of WS$_2$ sheets more enhanced the distribution kinetics and electrochemical activity. Phase control of the WS$_2$ is of important concern as this process can improve the Li-ion transfer. In details, the metallic (1T) stage of WS$_2$ has an octahedral synchronization that demonstrates equivalent characteristics and microstructure as associated to semiconducting (2H) phase of WS$_2$ like the trigonal prismatic orientation represented by the transversal movement of one of S atomic layer [31]. Conversely, limited studies have been documented concerning the preparation and use of 1T WS$_2$ in LIBs. Analogous to MoS$_2$, the fabrication path and approach of 1T WS$_2$ mostly comprise alkali metal and butyllithium intercalation, rare metal doping, pulsed laser deposition, and ammonia ions intercalation [31]. For instance, 1T-WS$_2$ nanoribbon was reported by Liu et al. [32] to have been produced via the intercalation of ammonia ions (NH_4^+) and the removal of 0.75 electrons per NH_4^+ from WS_2. In addition, electron supporter substitutes like Re, Tc and Mn can facilitate the phase transformation of WS_2 by forming covalent bonds resulting in a prominent escalation in the stable metallic WS_2. Nevertheless, some inadequacies tend to limit the request of these approaches; these include the complexity of the preparation process and the low scale production. Therefore, there are still channels for improving the electrochemical performances through choosing the efficient approach to achieve the 1T phase WS_2. Although WS_2 has received potential consideration for its remarkable capacity, ample resources as well as its low cost, still the dilemma on how to advance its cycle stability and reversible capability is quite a long way for research. Phase controlled synthesis of WS_2 is thus accelerated and studied for LIBs based on the phase purity of WS_2.

Wang et al. [31] stated the employment of metallic WS_2 (1T) phase material deposited on carbon fiber cloth for LIBs. This study revealed the transversal movement of S grains in WS_2 by nitrogen (N) atoms. In this manner, the cyclic strength of metallic phase WS_2 was achieved through N-W covalent bond formation. The enhanced electrochemical performance suggested that the introduction of metallic phase might not only diminish the preliminary discharge capability loss, then also could enhance the reversible capability compared to the conventional semi-conducting phase of WS_2 (2H).

A study conducted by Li et al. [33] demonstrated an in-situ synthesis procedure for WS_2 nanosheets deposited on compact graphene oxide hollow sphere (Hs-rGO) with N-doped graphene (NG) covering the outward layer to form NG-WS_2@Hs-rGO heterostructure. When employed as anode for LIBs, NG-WS_2@Hs-rGO heterostructure showed a great capacity of 1309 mAhg^{-1} at 100 mAg^{-1}, which delivers a release capacity of 301 mAhg^{-1} over 320 trails. The retention capacity extents as good as 106% of the release capacity at 10th cycle. It has been shown that the internal Hs-rGO delivers a space for the majority growth of WS_2, additionally the exterior layer of WS_2 deposited with NG also facilitates extra active edges for Li ion storage aside from enhancing the conductivity. Zhou et al. [34] employed electrospinned WS_2/Carbon nanofibers (i.e. WS_2/CNFs heterostrucutre) as anode materials for LIBs. This WS_2/CNFs heterostructure is composed of a single layer WS_2 consistently implanted in carbon fibers and exhibited discharge capacity of 756 mAhg^{-1} at 100 mA g^{-1}, a feature credited to its enormous specific surface area that accommodates large interfacial capacitance. The employed synthetic procedure revealed that the intrinsic flexible quality of electro-spun nanosized fibers rendered them an excessive probable in the deployment of binder free anodes. This procedure also is well suitable for additional comparable 2D layered materials aside from WS_2 and hence can provide a favorable aspirant electrode for bendable LIBs. Li et al. [35] described an in-situ synthesis of WS_2@rGO hybrid that could convey an improved capacity of 565 mAh g^{-1} at 100 mAg^{-1} after 100 sets with remarkable rate capacity of 337 $mAhg^{-1}$ at 2 Ag^{-1}. In other work, Liu et al. [36] demonstrated negative charged of WS_2/GO nanosheets in generating extremely effective flexible anode for LIBs assembled simply through a vacuum filtration process of homogeneous hybrid lamellar films. This additive free WS_2/rGO flakes electrode exhibited superior cycling steadiness with worthy rate capacity and a great reversible ability of 698 $mAhg^{-1}$ at 100 mAg^{-1} even after 100 cycles, which was significant larger than that of bare WS_2 and rGO nanosheets of 89 mAh g^{-1} and 60 $mAhg^{-1}$, respectively. This greater electrochemical behavior can be attributed to the consistently altered flaky structures, where rGO nanosheets diminished the restacking of WS_2 nanosheets, thereby improving the conductivity and endured the varying capacity in Li ion insertion-extraction reactions.

4.   MoSe₂ based anode materials for LIBs

Molybdenum diselenide ($MoSe_2$) is a one of the promising anode material for LIBs with narrow bandgap semiconductor. It consists of stacked layers of Se-Mo-Se sheet-like crystal structure creating it a good applicant for LIBs through electrochemical intercalation process. Unlike MoS_2. $MoSe_2$ might offer extra benefits because of the high electrical conductivity of Se as compare to S atoms [37]. With respect to their crystal orientation, $MoSe_2$ has two aspects: (i) semi-conductive trigonal 2H-$MoSe_2$ and (ii) fusible octahedral 1T-$MoSe_2$ phases, similar to MoS_2 and WS_2 materials. The 1T $MoSe_2$ construction has a great inherent electronic conductivity to defeat the compassion to air atmosphere by substituting a straight Li intercalation/exfoliation way with a NH_4^+ intercalation one. While, 2H-$MoSe_2$ typically agonizes from minor cycling stability and lower rate competence owing to its variability and inherent low electronic conductivity. To alter and direct the structure of $MoSe_2$, many strategies have been implemented such as nanostructure design, developing of integrated electrodes, and coupling with carbon-based conductive frameworks [38]. For instance, Mendoza-sanchez et al. [39] reported a liquid phase exfoliated $MoSe_2$ embedded to single-wall carbon nanotubes (SWCNTs) electrodes fabricated via a spray-assisted method and concurrently studied their electrochemical performance in a 1M $LiPF_6$-EC-DMC electrolyte in a standard three-probe cell within 3V electrochemical window. The cyclic stability of $MoSe_2$ is enhanced by SWCNTs. As fabricated $MoSe_2$-SWCNTs probes demonstrated a capability of 798 mAg^{-1} at a 0.05C rate for eight cycles, which is much greater compared to capability of a bulk $MoSe_2$-carbon black probes of 557 $mAhg^{-1}$ at 0.05C for first cycle. The $MoSe_2$/SWCNTs electrode showed capacity retaining of 58% over 100 sequences at a 0.5C rate. The enriched demonstration of the $MoSe_2$-SWCNTs probes was credited to the

Materials Research Forum LLC
https://doi.org/10.21741/9781644900918-3

nanosized $MoSe_2$ nanosized-platelets and the coupling effect with SWCNTs that facilitated mechanical strength. Furthermore, alike nano-replication approach for perforated TMDs was effectively realistic to develop other TMDs. It demonstrated a reversible capability above the theoretical reversible ability of $MoSe_2$ which is 422 mAh g^{-1} with respect to transformation reaction of Li ion tracked by making Mo elements in the insoluble Li_2Se matrix [40]. This is beneficial for improving the electrochemical performance especially for lamellar arranged materials, owing to the extra interfacial Li ion storage, consequently creating the bridge among batteries and supercapacitors at the nanoscale. Surface alteration or alteration of the band gap by establishment of heterojunction to improve the conductivity is more prominent strategy for proceeding sheet-like $MoSe_2$ as anode materials. Interestingly, nanosized $MoSe_2$ holds slower charge transfer conflict, restored electrochemical efficiency, and has better probable as anode materials associated to amorphous ones. However, they may existing a similar phase transition progression after the first cyclic charge-discharge course.

Wang et al. [41] demonstrated the conversion of MoO_3/PANI composite and Se powders in Ar gas atmosphere into nitrogen (N)-doped carbon anchored on a surface of $MoSe_2$ by simple heating. This synthetic method does not need a multi-step procedure or the application of extremely toxic reducing agents. This procedure showed the important role of PANI played in the assembly of N-fixed carbon@$MoSe_2$ through the reduction of MoO_3 and Se while being decayed and carbonized into N-doped carbon nanorods cores. The N-doped carbon@$MoSe_2$ demonstrated high discharging capability of 1275 mAhg^{-1} with a great reversible lithium drawing out capability of 928 mAhg^{-1} and a coulombic efficient of 72.8%. Even after 100 cyclic reactions, the N-doped carbon@$MoSe_2$ electrode delivered a reversible capability of 906 mAhg^{-1} with a capacity retention ratio of 97.6 %. This enhanced electrochemical performance can be pinpointed to the exclusive core/branch nanostructure of N-doped carbon@$MoSe_2$ and synergistic outcome between the N-doped carbon nanorods cores and $MoSe_2$ layer branches. In a different study, Su et al. [42] has employed an anode for LIBs developed from 1D $MoSe_2$/C nanorods hybrid derived from organic/inorganic precursors of $Mo_3O_{10}(C_2H_{10}N_2)$ with ensuing selenization and carbon coating process. The $MoSe_2$/C nanorods exhibited 835 mAhg^{-1} at 200 mAg^{-1} with retention of 755 mAhg^{-1} after 200 cycles. The remarkable electrochemical acts of the $MoSe_2$/C nanorod hybrid were due to the bicontinuous electron/ion pathways, small charge transfer resistance and robust structural features and stability. In the case of bare $MoSe_2$, the electrochemical reaction is limited attributable to a great volume expansion and little electronic conductivity. Therefore, the coupling of conductive materials with $MoSe_2$ demonstrated excellent electrochemical performance. Another study conducted by Liu et al. [43] demonstrated sheet-like $MoSe_2$/C hybrid and Zhu et al. [44] reported

$MoSe_2$ deposited on carbon fibers to form nanoheterojunction, both showed excellent reversible capacity with stable current densities. These results indicated that the electrochemical performance could be intensively enhanced by making the nano-heterostructures. However, these electrodes were demonstrated to have low packing density owing to their extraordinary specific surface area and large carbon weight percentages.

Cui et al. [45] developed an electrospinning strategy for coupling of graphene slips into carbon deposited $MoSe_2$ (i.e. $MoSe_2/C@G$ heterostructure) and subsequently employed as anode materials for LIBs. This heterostructure demonstrated superior discharge capability of 370 mAhg^{-1} when 200 cycles at 0.2 A g^{-1} with remarkable rate performance. Tang et al. [46] presented $MoSe_2$ microspheres consisting of unified nanoflakes that are covered with regular tinny N-doped carbon layer made via simplistic two-step approach of hydrothermal reaction assisted via dopamine thermal polymerization. This $MoSe_2/N$-carbon hybrid presented minor polarization resistance, restored reversibility with improved electrochemical functioning related to bare $MoSe_2$. This hybrid provided a good reversible capability of 689 mAh g^{-1} at 100 mAg^{-1} after 100 cycles with high rate capacity (698 mAhg^{-1} at 2 Ag^{-1}). This enhanced electrochemical functioning can be indorsed to the combination of N-fixed carbon layer and their ordered microsphere structure with extra active edge sites and rapid ionic/electronic transfer characteristics. Similarly, Wang et al. [47] has developed N-fixed carbon/$MoSe_2$ core/branch nanostructure by configuring a combination of polyaniline (PANI) inserted to MoO_3 nanorod composite and Se powder in argon environment. This hybrid demonstrated a extraordinary reversible capability of 906 mAhg^{-1}, and capacity retaining rate of 97.61% at 100 mAg^{-1} afterward 100 cycles and capacity rate of 560 mAh g^{-1} at 1 Ag^{-1}, owing to the exclusive structure of N-fixed carbon nanorod centers and the $MoSe_2$ nanosheet divisions.

To obtain the heterojunctions of TMDs, coupling qualities of various TMDs have been observed as a promising approach that alters their electronic properties and thereby posing a possibility of application in the conversion and storage energy. For instance, Yang et al. [48] developed a simple, low price and great yield solution-phase scheme for making MoS_2-$MoSe_2$ p-n heterojunctions with perpendicularly developed $MoSe_2$ through coating with amorphous carbon. This heterojunction exhibited an advanced and efficient electrochemical and Li-ion storage performance respectively, with a high reversible capability rate of 676 mAhg^{-1} 200 mA g^{-1} for 200 cycles. Recent studies showed that the formation of $MoS_{2(1-x)}Se_{2x}$ alloy through the addition of Se into MoS_2 facilitated the extension of the interlayer positioning for additional active diffusion of ions and electrons in the course of charge/discharge process. However, this has also reduced the energy barriers to permit over the layers because of the minor distortion and the diverged electric field in the basal levels as a result of the bigger radius exhibited by Se [49]. Wang et al. [49] synthesized an optimal 2D layered perforated $MoS_{2(1-x)}Se_x$/rGO hybrid and has shown to display a large capability of 830 mAhg^{-1} after 150 cycles at 100 mAg^{-1} and an improved capability of 415 mAhg^{-1} at 1000 mAg^{-1}. In addition, $MoSe_2$ can also be employed to improve the cycling and rate performance of TMCs. This has been substantiated by Jin et al. [50] who developed an electrode material of $Cu_{2-x}Se$@carbon nanosheets covered with $MoSe_2$ nanoparticles. This electrode material has exhibited an enriched reversible capability of 432 mAhg^{-1} after 100 cycles at 100 mAg^{-1} with improved rate capacity surpassing that of bare $Cu_{2-x}Se$@carbon (166 mAhg^{-1}) or pure phase $Cu_{2-x}Se$ nanosheets (82 mAh g^{-1}).

5. WSe_2-based anode materials for LIBs

Tungsten diselenide (WSe_2) is an energetic member of TMDs and has been regarded valuable due to its exclusive characteristics, having a narrow bandgap of 1.6 eV, enormous neighboring layer spacing of 0.648 nm which is two folds greater related to graphite, and also can provide an efficient reversible ion intercalation and separation. Further, WSe_2 has shown to have a good density and large volumetric capacity. The theoretical mass density of WSe_2 is 9.32 g cm^{-3} causing it to possess a good volumetric energy density which is considered extra prominent than the gravimetric energy density due to the inadequate battery packing space [51]. Several studies have proposed that WSe_2 might be a good choice for advanced lithium-ion anode materials. Yang et al. [52] documented the WSe_2 nanoplates as a probe material for both LIBs and SIBs, which demonstrated an established reversible capability and an extraordinary rate capability with long cyclic life of over 1500 cycles at 1000 mAg^{-1}. The charge-discharge process substantially revealed the reversible conversion mechanism shown as follows:

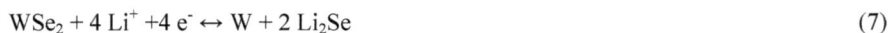

$$WSe_2 + 4\ Li^+ + 4\ e^- \leftrightarrow W + 2\ Li_2Se \tag{7}$$

Wang et al. [53] demonstrated an in-situ growth of hierarchical mesoporous nanoarchitecture of WSe_2 deposited on reduced graphene oxide (i.e. WSe_2/rGO) composite via facile one-step solvothermal method and consequently used as anode materials for LIBs. The WSe_2/rGO hybrid exhibited a large initial discharge capability of 744 mAhg^{-1} at 0.2C and maintained the capability of 528 mAhg^{-1} even after 80 cycles in contrast to a pure WSe_2 (549 mAhg^{-1}) which maintained 98 mAhg^{-1} after 80 cycles. This commendable electrochemical performance was attributed to its hierarchical structure.

rGO nanosheets provide more conductivity act as the skeleton of WSe_2/rGO anode providing the efficient electron transfer and also prevented the aggregation of WSe_2. Moreover, the rGO nanosheets also facilitated the in-situ growth of WSe_2 layers on the top of graphene with high active sites permitting an effective ionic diffusion and simple electrolyte infiltration.

6.   Other TMDs for LIBs

Many types of TMDs materials exhibited nanosheets morphology however, the relatively small electrical conductivity and weak structural stability in the electrochemical process hindered their prospective function as anode materials. In most cases, the nanosheets developed in the form of agglomerates mainly confer low electrochemical activity in the anodes due to the constrained specific surface area which affects the availability of Li ions. In order to overcome this demerit, many strategies have been taken into account such as the development of three dimensional (3D) flower-type micro/nanospheres, micro/nanostructures of 2D TMDs nanosheets, and coupling of TMD nanosheets in conductive carbonaceous and polymer-carriers with high certain surface areas.

It is well established that graphene nanosheets are a remarkable conductive substrate to dissolve and confine active materials due to its admirable conductivity, large surface area and strong mechanical strength. However, 2D graphene sheets could not restrict active anode materials excellently. Another mode of graphene (i.e. graphene nanoscrolls) comprising of several rolled-up graphene nanosheets has been demonstrated to demonstrate open-terminated tubular hollow structure which is reflected as a superior conductive substrate in confining active materials through providing open channels for more ion transport. Various sulphides such as TiS_2 and selenides (i.e. $TiSe_2$, VSe_2, and $NbSe_2$) were evaluated as cathodes for LIBs [54]. However, to obtain an acceptable energy packing, different required cathode features were sought: (i) large free energy of reaction; (ii) a variety composition ranges; (iii) a low change in free energy over configuration range n (mol Li/mol host); (iv) reversibility of reactions; (vi) in elevation diffusivity; (vii) indecipherability of the functioning materials in electrolyte medium; and (viii) good electronic conductivity. Among the TMDs explored, TiS_2 is superior for these criteria, which possesses an OCP of 2.47 V in comparison to Li metal with an power density of 480 $Whkg^{-1}$ in a half cell alignment [54]. However in the presence of oxygen, TiS_2 displayed more instability. Strikingly, VSe_2 demonstrated a greater rate of capacity (OCP of 2 V versus Li metal) than TiS_2 and has high stability and good reversibility.

Zhang et al. [55] demonstrated the thermal decomposition of graphene nanoscrolls linked by wrinkled graphene nanos ips as an efficient conductive complex for confirming tin disulfide (SnS_2) nanosheets for LIBs. During the thermal quenching, covering SnS_2

Materials Research Forum LLC
https://doi.org/10.21741/9781644900918-3

nanosheets in the nanoscolls formed a 2D/3D heterojunction. The ratio of nanoscolls and nanosheets can be achieved by merely adjusting the quenching conditions. The elevated $SnS_2/3DG$ possesses high specific surface area of 127.1 m^2g^{-1} with multi mesoporous which can efficiently hinder the aggregation of SnS_2 in order to facilitate an abundant sites for lithiation/delithiation. The $SnS_2/3DG$ hybrid demonstrated improved reversible capacity of 1514.8 $mAhg^{-1}$ at $01.Ag^{-1}$ with excellent rate capacity of 665.4 $mAhg^{-1}$ at 5 Ag^{-1} and good cyclic stability of $1050mAhg^{-1}$ at 50 cycles.

Table: 1 TMDs based nanocomposite electrode materials for lithium ion batteries

Electrode materials	Current density (mAg^{-1})	Capacity/number of cycles (mAhg^{-1})	Ref.
MoS$_2$ nanosheets	1000	900/1000	57
MoS$_2$ nanobelts	2000	480/100	58
MoS$_2$@3D vertical graphene	100	670/30	59
MoS$_2$@rGO nanosheets	100	420/160	60
MoS$_2$/C nanosheets	500	815/100	61
MoS$_2$/C nanospheres	100	523/100	62
MoS$_2$ on carbon monolayer	200	477/200	63
MoS$_2$/Fe$_3$O$_4$	2000	1033	64
MoS$_2$/TiO$_2$	100	710/100	65
MoS$_2$/Ni$_3$S$_2$	200	568	66
MoSe$_2$ nanosheets	1000	455/100	67
MoSe$_2$ microspheres	200	433/50	68
MoSe$_2$ nanoplates	42.2	380/50	69
MoSe$_2$ sheets/carbon	100	576.7/100	70
MoSe$_2$/Carbon nanospheres	1000	1208/150	71
N-doped carbon@MoSe$_2$	100	862.7	72

Table: 2 *Various TMDs materials coupled with graphene oxide electrode materials for lithium ion batteries.*

Materials	Key results	Ref
Layered MoS_2/rGO	Capacity of 750 mAhg^{-1} at 3 A g^{-1} and 1180 mAhg^{-1} after 80 cycles at 0.1 Ag^{-1}	73
MoS_2 nanoflakelet/rGO	Capacity of 1902 mAhg^{-1} for 1st discharging, high coulombic efficiency of 76.45%	74
MoS_2 nanowall/Gr	Reversible capacity of 700 mAhg^{-1} after 100 cycles at 0.05 Ag^{-1}	75
3D flower-like MoS_2/Gr	Reversible capacity of 980 mAhg^{-1} at 0.1 Ag^{-1} and 740 mAhg^{-1} after 100 cycles at 1 Ag^{-1}	76
Honeycomb-like MoS_2/Gr foam	Capacity of 1235.3 mAhg^{-1} at 0.2 Ag^{-1} and 85.8% retention after 60 cycles	77
Flower-like MoS_2/Gr/CNTs	Capacity of 300 mAhg^{-1} at 20 Ag^{-1} and 728 mAhg^{-1} after 1000 cycles at 5 Ag^{-1}	78
3D MoS_2/C_3N_4/N-rGO network	Capacity of 800 mAh g^{-1} at 0.1 Ag^{-1} and 91% retention after 100 cycles	79
3D $MoSe_2$/rGO foam	Capacity of 1160 mAh g^{-1} at 0.1 C and 650 mAh g^{-1} after 50 cycles at 0.5 C	80
Perpendicular $MoSe_2$/rGO	Capacity of 1001 mAh g^{-1} at 0.1 A g^{-1} and 725 mAh g^{-1} after 100 cycles	81
Ultrathin WS_2/3D Gr	Capacity of 766 mAh g^{-1} at 0.1 Ag^{-1} and 416 mAh g^{-1} after 100 cycles	82
Few-layered WS_2/rGO	Capacity of 400–450 mAh g^{-1} after 50 cycles at 0.1 A g^{-1}	83
Quasi-3D wrinkled $MoSe_2$/Gr	Reversible capacity of 1102 mAhg^{-1}	84
VSe_2/rGO	Capacity of 632 mAhg^{-1} at 0.1 A g^{-1} during 60 cycles	85
ReS_2/rGO	Capacity of 918 mAhg^{-1} at 0.2 C	86
V_5S_8/graphite nanosheet	Capacity of 1112 mAhg^{-1} at 0.1 Ag^{-1}	787 87

Cao et al. [56] reported the application of TiO_2-B@VS_2 heterojunction nanowire arrays as additive free anode for LIBs. This TiO_2-B@VS_2 heterojunction holds a superior reversible capability of 365 mAhg^{-1} over 500 cycles at a current density rate of 1C (335 mAg^{-1}). Moreover, TiO_2-B@VS_2 heterojunction has been shown to serve a high volume rate of 171 mAhg^{-1} at 10C rate assigned to the outstanding physical stability of the TiO_2 nanoarrays together with the high capacity and conductivity of VS_2. Various TMDs based nanohybrid for LIBs were tabulated in Table-1 and Table-2, reflecting the significane of 2D/2D heterojunction for power storage concerns.

7.   Summary and future outlooks

This chapter summarized the latest insights in the progress of graphene-type enveloped metal dichalcogenides such as MoS_2, WS_2, $MoSe_2$, and WSe_2 and their hybrids as conductor materials for LIBs. TMDs possess an extensive range of intercalation chemistry supported by the following reasons: (i) the amplified valence of the chalcogenide atoms facilitates the contribution of conversion metal valence, (d) paths in covalent attachment-this has the result of decreasing the recognized control on the metal center which fascinating metal-to-metal bond creation, and (ii) the greater diversity of the X_2^- anion permits for the acceptance of arrangements in which the surroundings of X is extremely irregular. All these materials have achieved prominent accomplishments up to date. However numerous challenges still exist that need to be solved in the future. Currently, the integration of TMDs with carbon medium such as graphene has shown to enhance the energy and power densities. This process however has reduced the initial coulombic efficiency. Furthermore, tamper with or intercalation via heteroatoms with S, Se, C is deemed promising for improving the conductivity as well as for extending interlayer spacing causing an enhancement in the electrochemical reactions. TMDs could be securely employed as anode materials exclusively for high power batteries due to the outstanding rate capability and higher cycling stability. There are still some important hurdles that need to be addressed in order for the commercial employment of progressive power packing gadgets based on the TMDs materials to be realized. It would be rational to trust that more important considerate and search in this emerging field would bridge us to new sensational outcomes for innovative high functioning rechargeable batteries. In order to optimize the LIBs performance, two major concerns should be taken into consideration (i) fundamental agreement the chemical reaction arising in TMD electrodes, and (ii) fabricating innovative TMDs devices.

Major issues concerning LIBs are (i) as commercial anodes for LIBs, graphite materials cannot encounter the continually-growing demands for high power density due to their limited specific capacity and (ii) unsatisfactory capability rates. The electrochemical endowment of TMDs at large current ditch is inhibited by their inadequate electronic conductivity regulating the reasonable power density. These materials have been exploited as novel class of anode material for LIBs exclusively for elevated rate uses in spite of their smaller theoretical capability. But, the studies on WS_2, WSe_2, and $MoSe_2$-based hybrids are still in their infancy. These materials possess limited cycling constancy due to enormous volume deviations through insertion/extraction progressions of ions. More progressive studies should give emphasis on the electrochemical functioning mechanism, the optimization of the electrode materials synthesis, and the control and manipulation of phase and structural stability. Especially, the procedure for the intercalation and conversion manners should be promote evaluated to elucidate the ongoing arguable in the transformation behavior of these transition metals and their reversible progression. Moreover, it is essential also to further established the capacity declining mechanism and impact of shuttle effect for TMDs. Indeed, more experimental and modeling studies on the electrochemical phenomena in high capability electrode materials for LIBs are need to be performed to completely understand the fundamental degradation of electrochemical performance during the cycling process. It has been scripted also in this chapter that the in-situ studies provided more evidences and concepts to overcome the electrochemical degradation phenomena. Additionally, it is very necessary to optimize and control the design schemes such as nanostructuring, nanoporosity, surface coating and compositing for justification of electrochemo-mechanical degradation.

Acknowledgement

This research work was supported by the National Research Foundation of Korea (NRF) (2020R1A2B5B01002744).

References

[1] Y. Gao, X. Wu, K. Huang, L. Xing, Y. Zhang, L. Liu, Two-dimensional transition metal diseleniums for energy storage application: A review of recent developments, Cryst. Eng. Comm. 19 (2017) 404-418. https://doi.org/10.1039/C6CE02223E

[2] J. Huang, Z. Wei, J. Liao, W. Ni, C. Wang, J. Ma, Molybdenum and tungsten chalcogenides for lithium/sodium-ion batteries: Beyond MoS_2, J. Energy Chem. 33 (2019) 100–124. https://doi.org/10.1016/j.jechem.2018.09.001

[3] T. Stephenson, Z. Li, B. Olsen, D. Mitlin, Lithium ion battery applications of molybdenum disulfide (MoS_2) nanocomposites, Energy Environ. Sci. 7 (2014) 209–231. https://doi.org/10.1039/C3EE42591F

[4] M. Pumera, Z. Sofer, A. Ambrosi, Layered transition metal dichalcogenides for electrochemical energy generation and storage, J. Mater. Chem. A 2 (2014) 8981–8987. https://doi.org/10.1039/C4TA00652F

[5] X. Fang, X. Guo, Y. Mao, C. Hua, L. Shen, Y. Hu, Z. Wang, F. Wu, L. Chen, Chem. Asian J. 7 (2012) 1013–1017. https://doi.org/10.1002/asia.201100796

[6] A.V. Murugan, M. Quintin, M.H. Delville, G. Campet, C.S. Gopinath, K. Vijayamohanan, Exfoliation-induced nanoribbon formation of poly(3,4-ethylene dioxythiophene) PEDOT between MoS_2 layers as cathode material for lithium batteries, J. Power Sources 156 (2006) 615–619. https://doi.org/10.1016/j.jpowsour.2005.06.022

[7] W. Wu, X.Y. Wang, X. Wang, S.Y. Yang, X.M. Liu, Q.Q. Chen, Effects of MoS_2 doping on the electrochemical performance of FeF_3 cathode materials for lithium-ion batteries, Mater. Lett. 63 (2009) 1788–1790. https://doi.org/10.1016/j.matlet.2009.05.041

[8] S. Wu, R. Xu, M. Lu, R. Ge, J. Iocozzia, C. Han, B. Jiang, Z. Lin, Graphene containing nanomaterials for lithium-ion batteries, Adv. Energy Mater. 5 (2015) 1500400. https://doi.org/10.1002/aenm.201500400

[9] M.U. Krishnan, M. Kaur, K. Singh, M. Kumar, A. Kumar, A synoptic review of MoS_2: Synthesis to applications, Superlatt. Microstruct. 128 (2019) 274-297. https://doi.org/10.1016/j.spmi.2019.02.005

[10] M.S. Das, M. Kim, J. Lee, W. Choi, Synthesis, Properties, and Applications of 2D Materials: A Comprehensive Review, Critical Rev. Solid State Mater. Sci. 39 (2014) 231-252. https://doi.org/10.1080/10408436.2013.836075

[11] X. Xie, S. Wang, K. Kretschmer, G. Wang, Two-dimensional layered compound based anode materials for lithium-ion batteries and sodium-ion batteries, J. Colloid Interface Sci. 499 (2017) 17–32. https://doi.org/10.1016/j.jcis.2017.03.077

[12] H. Yoo, A.P. Tiwari, J. Lee, D. Kim, J.H. Park, H. Lee, Cylindrical nanostructured MoS_2 directly grown on CNT composites for lithium-ion batteries, Nanoscale 7 (2015) 3404–3409. https://doi.org/10.1039/C4NR06348A

[13] Y.E. Miao, Y. Huang, L. Zhang, W. Fan, F. Lai, T. Liu, Electrospun porous carbon nanofiber@MoS_2 core/sheath fiber membranes as highly flexible and binder free anodes for lithium-ion batteries, Nanoscale 7 (2015) 11093–11101. https://doi.org/10.1039/C5NR02711J

[14] Z. Wan, J. Shao, J. Yun, H. Zheng, T. Gao, M. Shen, Q. Qu, H. Zheng, Core-shell structure of hierarchical quasi-hollow MoS_2 microspheres encapsulated porous carbon

as stable anode for Li-ion batteries, Small 10 (2014) 4975–4981.
https://doi.org/10.1002/smll.201401286

[15] Y. Teng, H. Zhao, Z. Zhang, Z. Li, Q. Xia, Y. Zhang, L. Zhao, X. Du, Z. Du, P.
Lv, K. Świerczek, MoS_2 Nanosheets Vertically Grown on Graphene Sheets for
Lithium-Ion Battery Anodes, ACS Nano 10 (2016) 8526-8535.
https://doi.org/10.1021/acsnano.6b03683

[16] Zhou, J. Qin, X. Zhang, C. Shi, E. Liu, J. Li, N. Zhao, C. He, 2D space-confined
synthesis of few-layer MoS_2 anchored on carbon nanosheet for lithium-ion battery
anode, ACS Nano 9 (2015) 3837–3848. https://doi.org/10.1021/nn506850e

[17] Z. Jian, B. Zhao, P. Liu, F. Li, M. Zheng, M. Chen, Y. Shi, H. Zhou, Fe_2O_3
nanocrystals anchored onto graphene nanosheets as the anode material for low-cost
sodium-ion batteries, Chem. Commun. 50 (2014) 1215–1217.
https://doi.org/10.1039/C3CC47977C

[18] M. Mao, L. Mei, D. Guo, L. Wu, D. Zhang, Q. Li, T. Wang, High electrochemical
performance based on the TiO_2 nanobelt@few-layered MoS_2 structure for lithium-ion
batteries, Nanoscale 6 (2014) 12350–12353. https://doi.org/10.1039/C4NR03991B

[19] X.Y. Yu, H. Hu, Y. Wang, H. Chen, X.W. Lou, Ultrathin MoS_2 nanosheets
supported on N-doped carbon nanoboxes with enhanced lithium storage and
electrocatalytic properties, Angew. Chem. Int. Ed. 54 (2015) 7395–7398.
https://doi.org/10.1002/anie.201502117

[20] S.H. Choi, Y.C. Kang, Enhanced Li^+ storage properties of few-layered MoS_2-C
composite microspheres embedded with Si nanopowder, Nano Res. 8 (2015) 2492–
2502. https://doi.org/10.1007/s12274-015-0757-3

[21] X.Shan, S. Zhang, N. Zhang, Y. CheN, H. Gao, X. Zhang, Synthesis and
characterization of three-dimensional MoS2@carbon fibers hierarchical architecture
with high capacity and high mass loading for Li-ion batteries, J. Colloid Interface Sci.
510 (2018) 327–333. https://doi.org/10.1016/j.jcis.2017.09.078

[22] H.Xue, S. Yue, J. Wang, Y. Zhao, Q. Li, M.Yin, S.Wang, C. Feng, Q.Wu, H. Li,
D. Shi, Q. Jiao, MoS_2 microsphere@ N-doped carbon composites as high
performance anode materials for lithium-ion batteries, J. Electroanal. Chem. 840
(2019) 230-236. https://doi.org/10.1016/j.jelechem.2019.03.058

[23] Z. Wang, M. Liu, G. Wei, P. Han, X. Zhao, J. Liu, Y. Zhou, J. Zhang, Hierarchical
self-supported C@TiO_2-MoS_2 core-shell nanofiber mats as flexible anode for
advanced lithium ion batteries, Appl. Surf. Sci. 423 (2017) 375–382.
https://doi.org/10.1016/j.apsusc.2017.06.129

[24]   G. Yang, X. Li, Y. Wang, Q. Li, Z. Yan, L. Cui, S. Sun, Y. Qu, H. Wang, Three-dimensional interconnected network few-layered MoS_2/N,S codoped graphene as anodes for enhanced reversible lithium and sodium storage, Electrochim. Acta 293 (2019) 47-59. https://doi.org/10.1016/j.electacta.2018.10.026

[25]   S. Wang, B. Liu, G. Zhi, X. Gong, Z. Gao, J. Zhang, Relaxing volume stress and promoting active sites in vertically grown 2D layered mesoporous $MoS_{2(1-x)}Se_{2x}$/rGO composites with enhanced capability and stability for lithium ion batteries, Electrochim. Acta 268 (2018) 424-434. https://doi.org/10.1016/j.electacta.2018.02.102

[26]   Q. Zhu, C. Zhao, Y. Bian, C. Mao, H. Peng, G. Li, K. Chen, MoS_2/nitrogen-doped carbon hybrid nanorods with expanded interlayer spacing as an advanced anode material for lithium ion batteries, Synth. Met. 235 (2018) 103–109. https://doi.org/10.1016/j.synthmet.2017.11.009

[27]   W. Yang, J. Wang, C. Si, Z. Peng, J. Frenzel, G. Eggeler, Z. Zhang, [001] preferentially-oriented 2D tungsten disulfide nanosheets as anode materials for superior lithium storage, J. Mater. Chem. A 3 (2015) 17811–17819. https://doi.org/10.1039/C5TA04176G

[28]   Y. Liu, W. Wang, H. Huang, L. Gu, Y. Wang, X. Peng, The highly enhanced performance of lamellar WS_2 nanosheet electrodes upon intercalation of single-walled carbon nanotubes for supercapacitors and lithium ions batteries, Chem. Commun. 50 (2014) 4485-4488. https://doi.org/10.1039/c4cc01622j

[29]   J. Zou, C. Liu , Z. Yang , C. Qi , X. Wang , Q. Qiao, X. Wu , T. Ren , Multilayer-cake WS_2/C nanocomposite as a high-performance anode material for lithium-ion batteries: "Regular" and "Alternate", ChemElectroChem 4 (2017) 2232–2236. https://doi.org/10.1002/celc.201700414

[30]   Y. Liu, N. Zhang, H. Kang, M. Shang, L. Jiao, J. Chen, WS_2 nanowires as a high performance anode for sodium-ion batteries, Chem. Eur. J. 21 (2015) 11878–11884. https://doi.org/10.1002/chem.201501759

[31]   T. Wang, C. Sun, M. Yang, L. Zhang, Y. Shao, Y. Wu, X. Hao, Enhanced reversible lithium ion storage in stable 1T@2H WS_2 nanosheet arrays anchored on carbon fiber, Electrochim. Acta 259 (2018) 1-8. https://doi.org/10.1016/j.electacta.2017.10.154

[32]   Q. Liu, X. Li, Z. Xiao, Y. Zhou, H. Chen, A. Khalil, T. Xiang, J. Xu, W. Chu, X. Wu, J. Yang, C. Wang, Y. Xiong, C. Jin, P.M. Ajayan, L. Song, Stable metallic 1T-WS_2 nanoribbons intercalated with ammonia ions: the correlation between structure

and electrical/optical properties, Adv. Mater. 27 (2015) 4837-4844.
https://doi.org/10.1002/adma.201502134

[33]   T.Li, R. Guo, Y. Luo, F. Li, Z. Liu, L. Meng, Z. Yang, H. Luo, Y. Wan, Innovative N-doped graphene-coated WS_2 nanosheets on graphene hollow spheres anode with double-sided protective structure for Li- Ion storage, Electrochim. Acta 290 (2018) 128-141. https://doi.org/10.1016/j.electacta.2018.09.065

[34]   S. Zhou, J. Chen, L. Gan, Q. Zhang, Z. Zheng, H. Li, T. Zhai, Scalable production of self-supported WS_2/CNFs by electrospinning as the anode for high-performance lithium-ion batteries, Sci. Bull. 61 (2016) 227–235. https://doi.org/10.1007/s11434-015-0992-8

[35]   H. Li, K. Yu, H. Fu , B. Guo , X. Lei , Z. Zhu , Multi-slice nanostructured WS_2@rGO with enhanced Li-ion battery performance and a comprehensive mechanistic investigation, Phys. Chem. Chem. Phys. 17 (2015) 29824–29833. https://doi.org/10.1039/C5CP04081G

[36]   Y. Liu, W. Wang, Y. Wang, X. Peng, Homogeneously assembling like-charged WS_2 and GO nanosheets lamellar composite films by filtration for highly efficient lithium ion batteries, Nano Energy 7 (2014) 25–32. https://doi.org/10.1016/j.nanoen.2014.04.018

[37]   J. Morales, J. Santos, J.L. Tirado, Electrochemical studies of lithium and sodium intercalation in $MoSe_2$, Solid State Ionics 83 (1996) 57–64. https://doi.org/10.1016/0167-2738(95)00234-0

[38]   J. Huang, Z. Wei, J. Liao, W. Ni, C. Wang, J. Ma, Molybdenum and tungsten chalcogenides for lithium/sodium-ion batteries: Beyond MoS_2, J. Energy Chem. 33 (2019) 100–124. https://doi.org/10.1016/j.jechem.2018.09.001

[39]   B. Mendoza-Sánchez, J. Coelho, A. Pokle, V. Nicolosi, A study of the charge storage properties of a $MoSe_2$ nanoplatelets/SWCNTs electrode in a Li-ion based electrolyte, Electrochim. Acta 192 (2016) 1–7. https://doi.org/10.1016/j.electacta.2016.01.114

[40]   L. Wu, P. Tan , Y. Liu , X. Xiong , J. Pan , Effects of Carbon Content on the Lithium-Storage Properties of $MoSe_2$-C Nanocomposites, ChemistrySelect 2 (2017) 8101–8107. https://doi.org/10.1002/slct.201700818

[41]   J. Wang, C. Peng, L. Zhang, Y. Fu, H. Li, X. Zhao, J. Zhu, X. Wang, Construction of N-doped carbon@$MoSe_2$ core/branch nanostructure via simultaneous formation of core and branch for high-performance lithium-ion batteries, Electrochim. Acta 256 (2017) 19–27. https://doi.org/10.1016/j.electacta.2017.09.129

[42]   Q. Su, X. Cao, X. Kong, Y. Wang, C. Peng, J. Chen, B. Yin, J. ShI, S. Liang, A. Pan, Carbon-encapsulated $MoSe_2/C$ nanorods derived from organic-inorganic hybrid enabling superior lithium/sodium storage, performances. Electrochimica Acta 292 (2018) 339-346. https://doi.org/10.1016/j.electacta.2018.09.154

[43]   Y. Liu, M. Zhu, D. Chen, Sheet-like $MoSe_2/C$ composites with enhanced Li-ion storage properties, J. Mater. Chem. A 3 (2015) 11857-11862. https://doi.org/10.1039/C5TA02100F

[44]   M. Zhu, Z. Luo, A. Pan, H. Yang, T. Zhu, S. Liang, G. Cao, N-doped one dimensional carbonaceous backbones supported $MoSe_2$ nanosheets as superior electrodes for energy storage and conversion, Chem. Eng. J. 334 (2018) 2190-2200. https://doi.org/10.1016/j.cej.2017.11.158

[45]   C. Cui, G. Zhou, W. Wei, L. Chen, C. Li, J. Yue, Boosting sodium-ion storage performance of $MoSe_2@C$ electrospinning nanofibers by embedding graphene nanosheets, J. Alloys Compound. 727 (2017) 1280-1287. https://doi.org/10.1016/j.jallcom.2017.08.258

[46]   W. Tang, D. Xie, T. Shen, X. Wang , D. Wang , X. Zhang , X. Xia , J. Wu , J. Tu, Construction of Nitrogen-Doped Carbon-Coated $MoSe_2$ Microspheres with Enhanced Performance for Lithium Storage, Chem. Eur. J. 23 (2017) 12924–12929. https://doi.org/10.1002/chem.201702840

[47]   J. Wang, C. Peng, L. Zhang , Y. Fu , H. Li , X. Zhao , J. Zhu , X. Wang, Construction of N-doped carbon@$MoSe_2$ core/branch nanostructure via simultaneous formation of core and branch for high-performance lithium-ion batteries, Electrochim. Acta 256 (2017) 19–27. https://doi.org/10.1016/j.electacta.2017.09.129

[48]   J. Yang, J. Zhu, J. Xu, C. Zhang, T. Liu, MoSe2 Nanosheet Array with Layered MoS2 Heterostructures for Superior Hydrogen Evolution and Lithium Storage Performance, ACS Appl. Mater. Interfaces 9 (2017) 44550–44559. https://doi.org/10.1021/acsami.7b15854

[49]   S. Wang , B. Liu , G. Zhi , X. Gong , Z. Gao , J. Zhang , Relaxing volume stress and promoting active sites in vertically grown 2D layered mesoporous $MoS_{2(1-x)}Se_{2x}$/rGO composites with enhanced capability and stability for lithium ion batteries, Electrochim. Acta 268 (2018) 424–434. https://doi.org/10.1016/j.electacta.2018.02.102

[50]   R. Jin, X. Liu, L. Yang, G. Li, S. Gao, Sandwich-like $Cu_{2-x}Se@C@MoSe_2$ nanosheets as an improved-performance anode for lithium-ion battery, Electrochim. Acta 259 (2018) 841–849. https://doi.org/10.1016/j.electacta.2017.11.044

[51]   X.Q. Wang, Y.F. Chen, B.J. Zheng, F. Qi, J.R. He, Q. Li, P.J. Li, W. L. Zhang, Graphene-like WSe2 nanosheets for efficient and stable hydrogen evolution. J. Alloys Compd. 691 (2017) 698–704. https://doi.org/10.1016/j.jallcom.2016.08.305

[52]   W. Yang, J. Wang, C. Si, Z. Peng, Z. Zhang, Tungsten diselenide nanoplates as advanced lithium/sodium ion electrode materials with different storage mechanisms, Nano Res. 10 (2017) 2584-2598. https://doi.org/10.1007/s12274-017-1460-3

[53]   X. Wang, J. He, B. Zheng, W Zhang, Y. Chen, Few-layered WSe$_2$ in-situ grown on graphene nanosheets as efficient anode for lithium-ion batteries, Electrochim. Acta 283 (2018) 1660-1667. https://doi.org/10.1016/j.electacta.2018.07.129

[54]   M.S. Whittingham, Chemistry of intercalation compounds: Metal guests in chalcogenide hosts, Progress Solid State Chem.12 (1) (1978)41–99. https://doi.org/10.1016/0079-6786(78)90003-1

[55]   Y. Zhang, C. Zhao, Z. Zeng, J. M. Ang, B. Che, Z. Wang, X. Lu, Graphene nanoscroll/nanosheet aerogels with confined SnS$_2$ nanosheets: simultaneous wrapping and bridging for high-performance lithium-ion battery anodes, Electrochim. Acta 278 (2018) 156-164. https://doi.org/10.1016/j.electacta.2018.05.031

[56]   M. Cao, L. Gao, X. Lv, Y. Shen, TiO$_2$-B@VS$_2$ heterogeneous nanowire arrays as superior anodes for lithium-ion batteries, J. Power Sources 350 (2017) 87-93. https://doi.org/10.1016/j.jpowsour.2017.03.070

[57]   Y.Liu, L. Zhang, Y. Zhao, T. Shen, X. Yan , C. Yu, H. Wang, H. Zeng, Novel plasma-engineered MoS$_2$ nanosheets for superior lithium-ion Batteries, J Alloys Compound. 787 (2019) 996-1003. https://doi.org/10.1016/j.jallcom.2019.02.156

[58]   Z. Zhang, S. Wu, J. Cheng, W. Zhang, MoS$_2$ nanobelts with (002) plane edges-enriched flat surfaces for high-rate sodium and lithium storage, Energ. Storage Mater. 15 (2018) 65–74. https://doi.org/10.1016/j.ensm.2018.03.013

[59]   B. Ouyang, Y. Wang, Z. Zhang, R.S. Rawat, MoS$_2$ anchored free-standing three dimensional vertical graphene foam based binder-free electrodes for enhanced lithium-ion storage, Electrochim. Acta 194 (2016) 151-160. https://doi.org/10.1016/j.electacta.2016.02.120

[60]   Z. Che, Y. Li, K. Chen, M. Wei, Hierarchical MoS$_2$@RGO nanosheets for high performance sodium storage, J. Power Sources 331 (2016) 50-57. https://doi.org/10.1016/j.jpowsour.2016.08.139

[61]   Q.-c. Pan, Y.-g. Huang, H.-q. Wang, G.-h. Yang, L.-c. Wang, J. Chen, Y.-h. Zan, Q.- y. Li, MoS$_2$/C nanosheets encapsulated Sn@SnO$_x$ nanoparticles as

highperformance lithium-iom battery anode material, Electrochim. Acta 197 (2016) 50-57. https://doi.org/10.1016/j.electacta.2016.03.051

[62]   B. Wang, Y. Xia, G. Wang, Y. Zhou, H. Wang, Core shell MoS_2/C nanospheres embedded in foam-like carbon sheets composite with an interconnected macroporous structure as stable and high-capacity anodes for sodium ion batteries, Chem. Eng. J. 309 (2017) 417-425. https://doi.org/10.1016/j.cej.2016.10.073

[63]   Z.-T. Shi, W. Kang, J. Xu, Y.-W. Sun, M. Jiang, T.-W. Ng, H.-T. Xue, D.Y.W. Yu, W. Zhang, C.-S. Lee, Hierarchical nanotubes assembled from MoS_2-carbon monolayer sandwiched superstructure nanosheets for high-performance sodium ion batteries, Nano Energy 22 (2016) 27-37. https://doi.org/10.1016/j.nanoen.2016.02.009

[64]   Y. Chen, B. Song, X. Tang, L. Lu, J. Xue, Ultrasmall Fe_3O_4 nanoparticle/MoS_2 nanosheet composites with superior performances for lithium ion batteries, Small 10 (2014) 1536–1543. https://doi.org/10.1002/smll.201302879

[65]   M. Mao, L. Mei, D. Guo, L. Wu, D. Zhang, Q. Li, T. Wang, High electrochemical performance based on the TiO_2 nanobelt@few-layered MoS_2 structure for lithium-ion batteries, Nanoscale 6 (2014) 12350–12353. https://doi.org/10.1039/C4NR03991B

[66]   J. Wang, J. Liu, H. Yang, D. Chao, J. Yan, S.V. Savilov, J. Lin, Z. X. Shen, MoS_2 nanosheets decorated Ni_3S_2@MoS_2 coaxial nanofibers: Constructing an ideal heterostructure for enhanced Na-ion storage, Nano Energy 20 (2016) 1–10. https://doi.org/10.1016/j.nanoen.2015.12.010

[67]   Y. Tang, Z. Zhao, Y. Wang, Y. Dong, Y. Liu, X. Wang, J. Qiu, Carbon-stabilized interlayer-expanded few-layer $MoSe_2$ nanosheets for sodium ion batteries with enhanced rate capability and cycling performance, ACS Appl. Mater. Interfaces 8 (2016) 32324-32332. https://doi.org/10.1021/acsami.6b11230

[68]   Y.N. Ko, S.H. Choi, S.B. Park, Y.C. Kang, Hierarchical $MoSe_2$ yolk-shell microspheres with superior Na-ion storage properties, Nanoscale 6 (2014) 10511-10515. https://doi.org/10.1039/C4NR02538E

[69]   H. Wang, X. Lan, D. Jiang, Y. Zhang, H. Zhong, Z. Zhang, Y. Jiang, Sodium storage and transport properties in pyrolysis synthesized $MoSe_2$ nanoplates for high performance sodium-ion batteries, J. Power Sources 283 (2015) 187-194. https://doi.org/10.1016/j.jpowsour.2015.02.096

[70]   Y. Liu, M. Zhu, D. Chen, Sheet-like $MoSe_2$/C composites with enhanced Li-ion storage properties, J. Mater. Chem. A 3 (2015) 11857-11862. https://doi.org/10.1039/C5TA02100F

[71] P. Ge, H. Hou, C. E. Banks, C. W. Foster, S. Li, Y. Zhang, J. He, C. Zhang, X. Ji, Binding MoSe$_2$ with carbon constrained in carbonous nanosphere towards high-capacity and ultrafast Li/Na-ion storage, Energy Storage Mater. 12 (2018) 310–323. https://doi.org/10.1016/j.ensm.2018.02.012

[72] M. Zhu, Z. Luo, A. Pan, H. Yanga, T. Zhu, S. Liang, G. Cao, N-doped one-dimensional carbonaceous backbones supported MoSe$_2$ nanosheets as superior electrodes for energy storage and conversion, Chem. Eng. J. 334 (2018) 2190–2200. https://doi.org/10.1016/j.cej.2017.11.158

[73] G. Liu, Y. Feng, Y. Li, M. Qin, H. An, W. Hu, et al. Three-dimensional multilayer assemblies of MoS$_2$/reduced graphene oxide for high-performance lithium ion batteries, Part Syst. Charact. 32 (2015) 489–497. https://doi.org/10.1002/ppsc.201400207

[74] J. Wang, X. Zhao, Y.Fu, X. Wang, A molybdenum disulfide/reduced oxide-graphene nanoflakelet-on-sheet structure for lithium ion batteries, Applied Surface Science 399 (2017) 237–244. https://doi.org/10.1016/j.apsusc.2016.12.029

[75] J. Guo, X. Chen, S. Jin, M. Zhang, C. Liang. Synthesis of graphene-like MoS$_2$ nanowall/graphene nanosheet hybrid materials with high lithium storage performance, Catal. Today 246 (2015) 165–171. https://doi.org/10.1016/j.cattod.2014.09.028

[76] L. Xu, Z. Jiao, P. Hu, Y. Wang, Y. Wang, H. Zhang. 3D MoS$_2$ nanoflowers decorated onto graphene nanosheets for high-performance lithium-ion batteries, Electrochem. Acta 3(9) (2016)1503–1512. https://doi.org/10.1002/celc.201600409

[77] J. Wang, J. Liu, D. Chao, J. Yan, J. Lin, ZX. Shen. Self-assembly of honeycomb-like MoS$_2$ nanoarchitectures anchored into graphene foam for enhanced lithium-ion storage, Adv. Mater .26 (2014) 7162–9. https://doi.org/10.1002/adma.201402728

[78] F. Pan, J. Wang, Z. Yang, L. Gu, Y. Yu MoS$_2$–graphene nanosheet–CNT hybrids with excellent electrochemical performances for lithium-ion batteries, RSC Adv. 5: 775 (2015) 18–26. https://doi.org/10.1039/C5RA13262B

[79] Y. Hou, J. Li, Z. Wen, S. Cui, C. Yuan, J. Chen, N-doped graphene/porous g-C$_3$N$_4$ nanosheets supported layered-MoS$_2$ hybrid as robust anode materials for lithium-ion batteries, Nano Energy 8 (2014) 157–64. https://doi.org/10.1016/j.nanoen.2014.06.003

[80] J. Yao, B. Liu, S. Ozden, J. Wu, S. Yang, M.T.F. Rodrigues, et al. 3D nanostructured molybdenum diselenide/graphene foam as anodes for long-cycle life lithium-ion batteries, Electrochim. Acta 176 (2015) 103–11. https://doi.org/10.1016/j.electacta.2015.06.138

[81]   Z. Luo, J. Zhou, L. Wang, G. Fang, A. Pan, S. Liang. Two-dimensional hybrid nanosheets of few layered $MoSe_2$ on reduced graphene oxide as anodes for long-cycle-life lithium-ion batteries, J. Mater. Chem. A; 4 (2016) 15302–8. https://doi.org/10.1039/C6TA04390A

[82]   G. Huang, H. Liu, S. Wang, X. Yang, B. Liu, H. Chen, et al. Hierarchical architecture of WS_2 nanosheets on graphene frameworks with enhanced electrochemical properties for lithium storage and hydrogen evolution, J. Mater. Chem. A 3 (2015) 24128–38. https://doi.org/10.1039/C5TA06840A

[83]   K. Shiva, H.S.S.R Matte, H.B. Rajendra, A.J. Bhattacharyya, C.N.R. Rao. Employing synergistic interactions between few-layer WS_2 and reduced graphene oxide to improve lithium storage, cyclability and rate capability of Li-ion batteries, Nano Energy 2 (2013)787–93. https://doi.org/10.1016/j.nanoen.2013.02.001

[84]   L. Ma, X. Zhou, L. Xu, X. Xu, L. Zhang, W. Chen. Ultrathin few-layered molybdenum selenide/graphene hybrid with superior electrochemical Li-storage performance, J. Power Sources 285 (2015) 274–80. https://doi.org/10.1016/j.jpowsour.2015.03.120

[85]   Y. Wang, B. Qian, H. Li, L. Liu, L. Chen, H. Jiang VSe_2/graphene nanocomposites as anode materials for lithium-ion batteries, Mater. Lett. 141 (2015) 35–8. https://doi.org/10.1016/j.matlet.2014.11.038

[86]   F. Qi, Y.Chen, B. Zheng, J.He, Q. Li, X. Wang, J. Lin, J. Zhou, B. Yu, P. Li, W. Zhang, Hierarchical architecture of ReS_2/rGO composites with enhanced electrochemical properties for lithium-ion batteries, Appl. Surf. Sci. 413 (2017) 123–128. https://doi.org/10.1016/j.apsusc.2017.03.296

[87]   X. Ou, X. Liang, F. Zheng, Q. Pan, J. Zhou, X. Xiong, C. Yang, R. Hu, M. Liu, Exfoliated V_5S_8/graphite nanosheet with excellent electrochemical performance for enhanced lithium storage, Chem. Eng. J. 320 (2017) 485–493. https://doi.org/10.1016/j.cej.2017.03.069

Lithium-ion Batteries - Materials and Applications
Materials Research Foundations 80 (2020) 91-122

Materials Research Forum LLC
https://doi.org/10.21741/9781644900918-4

Chapter 4

Metal Sulphides for Lithium-ion Batteries

Udaya Bhat K.[1*], Sunil Meti[1], C. Prabukumar[1], Suma Bhat[2]

[1] Department of Metallurgical and Materials Engineering, National Institute of Technology Karnataka, Surathkal, 575025, India

[2] Department of Mechanical Engineering, Srinivas School of Engineering, Mukka, 575025, India

*udayabhatk@gmail.com

Abstract

Increasing demand for flimsier, thinner, flexible batteries with higher capacities encourage present and future research for newer electrode materials with improved features. Metal sulphide based nanomaterials, due to their unique properties have opened a new domain for exploration in the domain of lithium-ion batteries. This review summarizes various metal sulphides, their developmental highlights and opportunities as the electrode materials. It offers updated knowledge on various metal sulphide systems. Review concludes by highlighting the promises metal sulphides as electrodes for the future lithium-ion batteries and challenges to be crossed to make it successful.

Keywords

Lithium-ion Batteries, Metal Sulphides, Improved Performance, Nanocomposites

Contents

1. Introduction

Increased awareness of the modern society on environmental pollution and dependence on the uninterrupted and portable power sources drives the focus on sustainable, efficient, economical energy production and its management. Electrical energy storage devices, like batteries and capacitors play pivotal part in the management of environmentally friendly sustainable energy [1]. Typical energy storage technologies for portable power devices are fuel cells, batteries, pseudocapacitors, etc. They all work on the basis of conversion of chemical potential into electricity [2].

Presently, batteries are the main source for powering devices which are both stationary and mobile in nature [3]. Rechargeable batteries (or secondary batteries), such as lithium-ion batteries (LIBs), metal-air batteries, Na-ion batteries, Mg ion batteries are all devices working on the repeated reversible conversions of energy from chemical to electrical ones. Among various secondary batteries, LIBs are considered to be the most important energy storage and conversion units [4,5].

2. Demands on batteries in 21st Century

The first commercialized batteries are introduced by Sony to the market. Since then, the rechargeable LIBs have made a niche market share quickly [6]. The LIBs are already been used as the main source of energy due to their relatively high energy density, capacity, and power density, efficient cycling stability and unimpeachable power performance [7]. The global demand of growing power consumption has stimulated the world to look for the next generation Li batteries with efficient storage technologies for supporting low carbon society. The next-generation LIBs are future of energy storage devices in the main application of hybrid electric vehicles and smart grids. Presently, the researchers are mainly focused on exploiting new flexible, lighter and thin electrode materials to improve energy density for LIBs [8]. The growing consumption of non-renewable fossil fuels and environmental pollution has attracted the researchers to explore the eco-friendly renewable energy sources, like nuclear power plants, solar and wind. In this context the LIBs system play a pivotal role in the energy management which can be harvested from the renewable energy sources [9].

3. Design of a lithium-ion battery (LIB)

Positive electrode (cathode), negative electrode (anode), separator and liquid electrolyte (ion medium) are the four important components of LIBs (Fig. 1). The electrolyte conductor medium helps in the transportation of ions between the electrodes. The electrolyte should provide the environment for ions to diffuse with high mobility so that the active materials (cathode and anode) will be efficiently utilized. Electrolytes can be organic or aqueous. It is well accepted that organic lithium-ion battery has matchless performance, flexible in design, and excellent energy density. They show characteristic features, like longer lifetime, superior energy capacity, and weightless in comparison with other secondary batteries which were in use [10,11]. Though LIBs with organic electrolytes show excellent performance, aqueous rechargeable LIBs offer the benefit of lower cost, higher safety and environmentally benign nature [10,11].

In LIB, the ion transfer reaction occurs between two active materials, anode and cathode. In a real system, many side reactions also occur. In general, the mechanism of the active materials in the aqueous electrolyte is complex compared to that in the organic material.

During discharge cycle in LIBs, the lithium ions transfer from the negative electrode to positive electrode and is reversed in charging cycle. Lithium intercalation compounds, such as $LiCoO_2$, $LiMnO_4$, $LiFeO_4$, $Li_3V_2(PO_4)_3$ are the commonly used cathode materials in LIBs. In LIBs, graphite and transition metal oxides are the major choices for the anode material. Amongst these, many anode materials show good cyclic performance but low initial charge-discharge efficiency. Transition metal oxides have good cyclic efficiency but highly irreversible capacity loss in the first cycle [12]. One of the potential solutions to this problem is to develop new electrode materials. Improved electrodes allow for the storage of more lithium ions and increase the battery capacity [13].

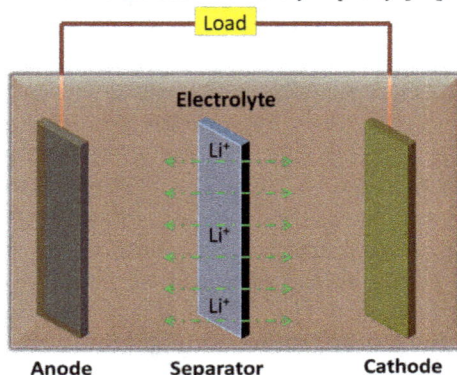

Figure 1. Schematic configuration of LIBs.

4.   Materials related issues in LIBs in modern era

The morphology, composition and microstructure of the electrode materials are crucial for the efficient electrochemical performance of the batteries [14,15]. The characteristic electrode materials should have the properties, like high conductivity, increased surface area, and short traveling distance for lithium ions. They could be tailored through the materials processing route. Further, the use of the nanostructured materials obtained a special recognition as they offer unique features like short diffusion lengths, high surface area and increased electronic/ionic conductivity. These features facilitates easy intercalation of Li-ions into/from the host electrodes in very short time [16,17].

The electrochemical batteries performance is mostly dependent on the electrode materials. At present, commercial graphite is the predominant choice as the anode material because of its high Columbic efficiency and flat potential profile. However, due to graphite's low theoretical capacity of 372 mAh g^{-1}, poor rate capability and safety problems limit its applicability for high energy density targets. The low power density of LIBs is due to the slow diffusion (10^{-8} cm^{2}s^{-1}) of lithium ions in the graphite structure. The graphite anode also has a major drawback of forming lithium dendrites which limit its applicability in the high-power electric devices (like vehicles). Added to that the production cost and safety concern, especially in the mass-energy storage system is high

Materials Research Forum LLC
https://doi.org/10.21741/9781644900918-4

[11]. This has forced the researchers to explore and develop new-type anode materials [18]. These challenges could be overcome by using numerous candidates, like metals, semimetals, metal sulphides and metal oxides. Amongst these, metal sulphides are considered as the best alternate candidates for the anode material over the graphite which offers the advantage of large capacity.

It is already mentioned that charge/discharge rates of LIBs is completely dependent on the mobility of the lithium ions and the electrons through the electrolyte and the active electrode material. Various strategies have been followed to increase the lithium-ion and electron transport kinetics. Also, the structural features promoting high ion and electron transport [19,20] or by reducing the path length over which electrons and lithium ions have to move in the batteries. With these strategies, a series of three-dimensional hybrid electrode materials [20], which can provide high charge–discharge rates have been developed.

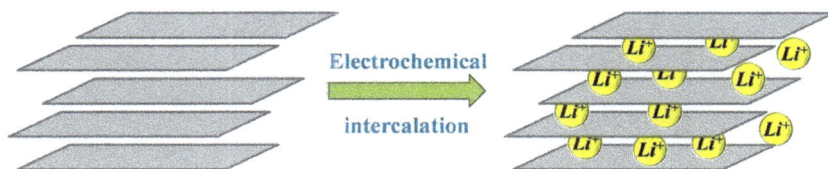

Figure 2. Intercalation of Li-ions into the layers of metal sulphides.

5.   Advantages of metal-sulphides for LIBs

Metal sulphides are bound to exhibit less volumetric expansion upon lithiation compared to metal oxides. Among various metal sulphides, transition metal dichalcogenides (TMDs) are highly promising because of their layered structure (Fig. 2). Interlayer spacing of TMDs provide a scope for the accommodation of species, like lithium ions [21]. Layered metal sulphides have the potential application as anode materials in LIBs due its various advantages such as better reversible discharge capacity and cycling performance compared to the metal oxides. The exfoliated few-layered metal sulphides exhibits different chemical and physical properties compared to their bulk counterparts. A large interlamellar spacing, with dimension between two layers ranging in nanometers accommodates volumetric expansion upon lithium storage. Few-layered metal sulphides can provide high volume ion diffusion paths and higher number of active sites for lithium-ion storage.

A layered metal sulphide has the potential to be used as high-power anode material due to the fast lithium ions storage kinetics resulting from surface pseudocapacitive storage behaviour. In addition, tailored few-layered electrodes are proficient members of being accommodated at high current densities. The metal sulphide anodes prepared by the intercalation are accommodating lithium ions through conversion reaction mechanism (Fig. 2). This mechanism offers higher energy density than intercalated metal sulphides. The mechanism of conversion reaction to store lithium ions during charging-discharging is shown by the equations below.

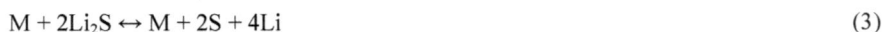

$$MS_2 + xLi \rightarrow Li_xMS_2 \tag{1}$$

$$Li_xMS_2 + (4 - x)Li \rightarrow M + 2Li_2S \tag{2}$$

$$M + 2Li_2S \leftrightarrow M + 2S + 4Li \tag{3}$$

In the first discharge cycle, the Li^+ ions are intercalated into MS_2 (M=Mo, W, Sn, Fe) to form Li_xMS_2. Subsequently, the Li_xMS_2 is decomposed into M and Li_2S. During charging cycle, a reversible reaction of Li_2S converting into S and Li occurs with M remaining unaffected [22].

Nevertheless, the main limitation of conversion reaction mechanism is that it may bring a large volume expansion of electrode material during lithiation/delithiation in the initial cycle. These factors results in mechanical failure of electrode materials by means of cracking, fracture and electrode disconnection from the current collectors which results in speedy fading of LIBs capacity. Mainly, the transition metal sulphides (TMSs) have drawn the attention as the anode material in LIBs due to their increased specific capacity, cheap, highly safe and environmental-friendly. The TMSs are resistant to formation of the lithium dendrites compared with commercial graphite anodes. This is due to the high working voltage platform of TMSs (above 1.0 V). With all these salient features, the TMSs are competitive enough to be used as the next-generation anode materials.

6. Metal sulphide based nanocomposites for battery applications

Conventionally metal dichalcogenides (metal sulphides) are observed as competitive materials in the applications of an anode electrode in LIBs. The limitations could be limited electrical conductivity along with capacity loss during the first cycle. Using combinations of heterogenous nanostructures can overcome these limitations [23]. Use of the sulphide/rGO or such composite makes the electrodes a highly flexible and opens up new opportunities, like roll up devices, wearable devices, health monitoring, skin devices, etc. These devices needs flexible power sources with high capacity and rate performance which will allow them to be used for fairly long time and to be charged at a short time [20,24]. A design configuration of metal dichalcogenide/rGO is found to be highly beneficial [25]. It showed enhanced performance and delivered reversible expulsion capability. Also, adding metal chalcogenide with rGO will alleviate volume changes during charge –discharge cycles [26].

7. Different types of metal sulphides as anode materials in the LIBs applications

Transition metal dichalcogenides (TMDs) are candidates for recharging batteries. Several metal oxides of the type MS_x (M= Cu, Co, Zn, Ti, Sn, Mo, W, Ni, etc) are explored for the possible application as the anode material in LIBs. Various methods are employed to evaluate the cyclic stability of the metal sulphides in rechargeable LIBs. The metal sulphide nanostructures and their composites integrated with carbonaceous materials (like graphene, CNT) have shown significant improved performance on cyclic performance, several challenges still remain. In the initial lithiation cycle, the MS_x structures are destroyed by converting to metal (M) and Li_2S. These initial structures, scales of the MS_x structures, the microstructure and dispersion characteristics of metal and LiS_2 in the electrodes are the factors affecting the cyclic stability of LIBs. The lithiation of MS_x results in massive volume expansion, leading to grinding (pulverization) and deterioration of electrode materials. This effect causes MS_x to result in poor ionic/electronic conductivity hampering their high-power performances.

7.1 Layered metal-sulphides for LIBs.

From the literature is seen that the layered structures have shown better performance compared the other materials. MoS_2, WS_2, ZrS_2, VS_2, SnS_2 are some of the layered (sulphide-metal-sulphides) materials which are similar to the graphite structure. These layered structures are two dimensional sandwiched sheets which are held by weak Van der Waals force. The layered structure of MoS_2 and WS_2 are similar where Mo (or W) is connected to six sulfur atoms. These structures are interconnected to give two-dimensional (2D) layered material, as shown in Fig. 3. ZrS_2 and VS_2 possess hexagonal close-packed lattice structure where Zr (or V) occupies the octahedral position in alternating layers and sulfur occupies hexagonal position. VS_4 is also layered structured linear chain compound similar to VS_2 with alternating bonding and non-bonding contacts between octal-coordinated V centers. Each sulfur unit in the structure bridges the two neighboring V atoms. SnS_2 is also a layered structure with CdI_2-type crystallographic

Lithium-ion Batteries - Materials and Applications   Materials Research Forum LLC
Materials Research Foundations **80** (2020) 91-122   https://doi.org/10.21741/9781644900918-4

characteristics. The sandwiched structure of SnS_2 contains Sn atoms between the two layers of sulfur atoms.

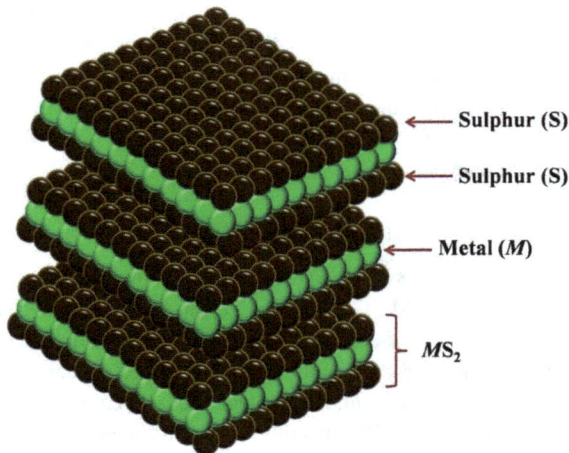

Figure 3. Schematic representation of layered metal sulphide structure.

7.2   Copper sulphides

The reaction involving copper sulphides with lithium offer an improved high capacity and energy density. Hence, the copper sulphides are the best TMSs. However, the reversibility of the copper sulphides is poor and hence nanostructuring is necessary to improve the properties. CuS and Cu_2S are the main sulphide forms of the copper. Based on the conversion reaction ($CuS + 2Li^+ + 2e^- \leftrightarrow Li_2S + Cu$), the specific capacity and electrical conductivity are calculated to be ~560 mAh g^{-1} and ~10^4 Scm^{-1}, respectively [27,28]. However, the Cu_2S is electrochemically more stable compared to the CuS [27]. In Cu_2S, the cell volume and ionic radius of Cu^+ ion (0.077 nm) is very close to the Li^+ ion and Li_2S (fcc). The volume expansion of Cu_2S is comparable to Li_2S during charge-discharge cycles. Jache and co-workers have predicted the mechanism of reaction for Cu_2S where it is displacement reaction than conventional conversion reaction [29]. The ex-situ techniques, such as X-ray diffractometry (XRD) and transmission electron microscopy (TEM) are employed to study the phase transformation and electrochemical reactivity of copper sulphides during the lithiation-delithiation cycles [30]. This TEM analysis also showed the extrusion of copper when cell is fully discharged. This unique electrochemical behaviour is present in Cu_2S compared to other sulphide counterparts

[31]. The addition of excess amount of copper to the copper sulphides electrodes will significantly improve the electrochemical performance and cycling stability [32].

The reaction between Cu_2S with Li is displacement reaction rather than conversion in the copper nanodomains. One of the interesting phenomenon in copper sulphides is the formation of macroscopic metallic Cu dendrites upon reduction [30,33]. The large copper dendrites will surround the Li_2S matrix after it is fully discharged [34]. It is already discussed that the Li_2S, Cu_2S and CuS have structural similarity (like space group and cell volume) accounting for the efficient electrochemical performances of the copper sulphides. The biggest disadvantage in copper sulphides system is the selection of narrow voltage window between 1.0 V to 3.0 V. This voltage range causes low releasing capacity during cycling compared with the voltage range of 0.01 and 3.0 V [35,36]. Usage of nano-metal sulphides and complexing are the two kinds of strategies followed to overcome the demerits of volume expansion and electronic insulating property of the discharge product (Li_2S). However, Nanomaterials effectively address the volume expansion and hence they are adopted for the synthesis of anode material [28].

Cu_2S nanowire arrays were synthesized by following the solution synthesis method [37]. These Cu_2S nanowire arrays are grown directly onto the copper metal current collector. As discussed, nanostructures exhibit stable Li^+ ion movement and increased reversible lithium storage capability without forming the copper dendrites. The electrochemical performance of Cu_2S nanowire array was tested with voltage range of 0.05 – 3.0 V, at room temperature. The first discharge curve had one voltage plateaus at 1.60 V and showed a high discharge capacity of 470 mAh g^{-1}. The first charge curve had two voltage plateaus at 1.85 V and 2.25 V with charge capacity of 400 mAh g^{-1}. The specific capacity and the Coulombic efficiency of Cu_2S was calculated to be 185 mAh g^{-1} and 96 %, respectively. The specific capacity after 85 cycles was found to be 79 %. This value lasts up to 100th cycle. Feng et al. synthesized reduced graphene oxide (rGO) incorporated copper sulphide nanowires [38]. The rGO/CuS showed the reversible capacity of 620 mAh g^{-1} at 0.5 C (1 C = 560 mAh g^{-1}). After 100 cycles it decreased to 320 mAh g^{-1} at current rate of 4 C. The synergetic effect between the rGO nanosheets and CuS nanowires resulted in excellent lithium storage performance. Feng et al. synthesized CuS nanowires with dimension of 6 nm diameter by template and surfactant-free method using dimethyl sulfoxide (DMSO)-ethyl glycol (EG) mixed solvent [39]. At 0.2 C, the initial releasing capacity and charge capacity are 832 mAh g^{-1} and 518 mAh g^{-1}, respectively. The charge capacity of 196 mAh g^{-1} and Coulombic efficiency of 99.0 % are achieved when the current is increased to 4 C. Jache et al. [29] reported charge capability of the copper sulphide electrode with variation in electrolyte composition. Also, it is shown that the Cu_2S could be recycled 150 times with Columbic efficiency of over 98.0 %, with overvoltage of 200 mV at different current rates (up to 1 C). It is also seen from the work that the performance of nanostructured copper sulphides is better compared to the bulk counterpart.

7.3 Cobalt sulphides

Cobalt sulphides (CoS_x) exhibit excellent physical and chemical properties. Various stoichiometric cobalt sulphides (such as CoS, CoS_2, Co_3S_4 and Co_9S_8) have been explored widely for the application as electrode material in LIBs. The CoS_x exhibits many interesting features, like natural abundance, high electrical conductivity and high theoretical capacities (CoS: 589 mAh g^{-1}, CoS_2: 870 mAh g^{-1}, Co_3S_4: 702 mAh g^{-1} and Co_9S_8: 545 mAh g^{-1}) [40]. The crystal structure of CoS, CoS_2, Co_3S_4 and Co_9S_8 is hexagonal, cubic, spinel and cubic close-packed arrangements, respectively [22]. Among various cobalt sulphides, Co_9S_8 is highly explored in the application of LIBs. Various carbonaceous nano materials are used to prepare Co_9S_8 composites to improve the conductivity and electrochemical activity. Many carbon nanostructures are used to prepare various Co_9S_8 nanocomposites, such as CNT/ Co_9S_8 [41], graphene nanosheets/ Co_9S_8 [40], porous carbon nanofibers/Co_9S_8 [42] and carbon nanotube aerogel/Co_9S_8 [43]. These carbonaceous structures (like CNT, GO, etc) inhibit the aggregation of Co_9S_8 by providing higher number of electrochemical active sites during cycling of LIBs. Li *et al.* [40] synthesized rGO/cobalt sulphide composite by following ultrasound-assisted wet chemical method. The morphological analysis depicts that the cobalt sulphide nanoparticles are consistently attached to rGO nanosheets. The electrochemical study on rGO/cobalt sulphide nanoparticles shows the high reversible capacity of 994 mAh g^{-1} after 150 cycles at a current density of 200 mA g^{-1}. The nanocomposite of rGO and cobalt sulphides shows the superior electrochemical performance over pure cobalt sulphide. Yongsheng *et al.* [44] developed a new cobalt sulphide encapsulated nanowire composite within nitrogen-doped porous branched carbon nanotubes (NBNTs) for LIBs. The NBNTs showed the reversible specific capacity of 1310 mAh g^{-1}, at a current density of 0.1 A g^{-1} with Columbic efficiency of almost 100 % for 200 cycles. This outstanding rate and cycling capability is due to the one dimensional porous Co_9S_8/NBNTs inter networks. CNT confinement and nitrogen doping of cobalt sulphide nanowires offers narrow electron pathway for individual nanoparticle and protecting the cobalt sulphide nanowires from pulverization.

Zhang *et al.* [43] worked on synthesizing three-dimensional (3D) aerogel consisting of CNT, graphene nanosheets and CoS_2 nanoparticles as electrode material for LIBs. The nanocomposite exhibited high reversible capacity of 975 mAh g^{-1} after 100 cycles at current density of 0.25 A g^{-1}. The high capacity is due to the synergetic effects of individual components. Liu *et al.* [45] synthesized the hollow cobalt sulphide nanoparticles embedded in graphitic carbon nanocages by following top-down approach. The composite showed superior lithium ion storage capability at working voltage range of 1.0 V to 3.0 V and displaying higher energy density of 707 Wh kg^{-1}. The reversible rate capacity shown by the composite is 536 mAh g^{-1} at current density of 0.2 C and 278 mAh g^{-1} at 10 C. when the working voltage is increased from 0.01 V to 3.0 V the capacity is found out to be 1600 mAh g^{-1} at current density of 100 mA g^{-1}. Chen *et al.* [46] worked on the synthesis of hollow microspheres of cobalt sulphide (CoS_2). The CoS_2 showed superior rate capacity of 720 mAh g^{-1} after 200 cycles. Jin *et al.* [47] synthesized worm-like CoS_2 assembled by ultrathin nanosheets with an average thickness of 2.1 nm by using solvothermal technique, without the assistance of surfactant or template. The CoS_2 worm-like structure showed discharge capacity of 883 mAh g^{-1} after 100 cycles at a current density of 100 mA g^{-1}. The material also shows higher rate capacity of 501 mAh g^{-1} at higher current density of 2000 mA g^{-1} after 50 cycles. Hu *et al.* [48] worked on the synthesis of binder-free and self-standing cobalt sulphide encapsulated in CNT and it showed extraordinary cycle stability and maintains 87 mAh g^{-1} after 6000 cycles at 1 A g^{-1}. The CNT/cobalt sulphide composite exhibited discharge capacity of 315 mAh g^{-1} at 100 mA g^{-1}. Zhou *et al.* [49] have developed mesoporous Co_9S_8 by solvothermal method and subsequent high temperature annealing. The Co_9S_8 hollow nanoparticles have shown high reversible rate capacity of ~1414 mAh g^{-1} after 100 cycles at 100 mA g^{-1}. The growth of carbon shell on hollow Co_9S_8 nanoparticles has improved the reversible capacity of ~896 mAh g^{-1} after 800 cycles at 2 A g^{-1}.

Researchers have also designed different structures of cobalt sulphides and also the composites of cobalt sulphides to overcome the limitations of the bare cobalt sulphide. The special architectures have shown promising performances for high discharge capacity and better cycling rate in LIBs. These well-defined large surface area structures of cobalt sulphides could largely reduce the diffusion paths of Li$^+$ ions and electrons. These features of cobalt sulphide nanostructures could result in high lithium ion storage capacity and easy lithiation-delithiation conversion reactions.

7.4 Molybdenum disulphide (MoS_2)

MoS_2 is a two-dimensional layered material where a layer of molybdenum atoms is sandwiched between the layers of sulphur atoms. This can be prepared like a sheet consisting of single to few layers of MoS_2 to offer high surface area [50]. There are different synthesis techniques to prepare the MoS_2 with different nanostructures.

Tian *et al.* synthesised MoS_2 nanospheres by following hydrothermal method [51]. Ammonium molybdate ($(NH_4)_6Mo_7C_{24} \cdot 4H_2O$) and sodium sulphide ($Na_2S \cdot 9H_2O$) were used as the source for molybdenum and sulphur, respectively. The produced material was MoS_2 nanospheres with the average diameter of 30 nm. Park *et al.* synthesised MoS_2 nanospheres using sodium molybdate dihydrate and L-cysteine as the source method by using hydrothermal method [52]. The galvanostatic study performed on this prepared nanospheres has shown that the MoS_2 nanospheres exhibited the charge capacity of 706 mAh g^{-1} at 100 mA g^{-1} and 658.1 mAh g^{-1} at 1000 mA g^{-1} after 30 number of cycles. This high capacity value was attributed to its nanostructure having increased interplanar distance and associated faster diffusion of lithium ions [52]. Guo *et al.* synthesized yolk-shell MoS_2 nanospheres with carbon shell [53]. The material showed the capacity value of 1065 mAh g^{-1} at the current density of 0.1 A g^{-1}. The electrochemical performance of the material was excellent as it exhibited the high capacity of 947, 914 and 847 mAh g^{-1} at high current density of 1, 2 and 5 A g^{-1}, respectively. The reason for this excellent performance was the result of the high stability and reversibility provided by the porous nanostructures and carbon shell [53].

MoS_2 nanosheets are another significant nanostructured material. This could be prepared by many techniques, such as hydrothermal [54], chemical vapour deposition (CVD) [55] and solvent assisted exfoliation [56]. The MoS_2 nanosheets were prepared by our group via solvent assisted exfoliation route by ultrasonication process [57]. Some details are given in Fig. 4 and Fig. 5. A TEM micrograph is shown in Fig.5a. It is observed that the prepared nanosheets contain only a few layers of the MoS_2. The study of XRD pattern (in Fig. 5b) shows that the peaks belonging to the planes (103) and (105) were absent in the exfoliated nanosheets. Liu *et al.* synthesized MoS_2 nanosheets by following hydrothermal method [58]. Subsequently, the prepared nanosheets were treated by oxygen plasma in order to be used in lithium-ion batteries. Then the prepared MoS_2 nanosheets were used as the working electrode in a coin-type battery. The charge-discharge studies exhibited the charge capacity of 1120-1180 mAh g^{-1} at 1 A g^{-1} and retained up to 600 cycles. It retained the capacity in the range of 920-1150 mAh g^{-1} at 1 A g^{-1} even after 1000 number of cycles. This high capacity retention of the MoS_2 nanosheets was attributed to the expansion of interlayer in MoS_2 and plasma-treatment induced defects that acted as the extra active sites for the Li-ions intercalation [58]. Xiang *et al.* worked on growing vertical MoS_2 nanosheets on the surface of the graphene [59]. The metallic $1T-MoS_2$ nanosheets on the graphene were conductive to facilitate the charge transportation. The calculated charge capacity of the material was 666 mAh g^{-1} at a current density of 3.5 A g^{-1}.

There are many investigations on the exploration of the MoS_2 based nanocomposite in the battery domain. MoS_2/graphene nanocomposites were used as the anodes. Exfoliated MoS_2/rGO composite could afford a capacity of 165 mAh g^{-1} after 50 cycles [60]. Qin *et*

Lithium-ion Batteries - Materials and Applications   Materials Research Forum LLC
Materials Research Foundations **80** (2020) 91-122   https://doi.org/10.21741/9781644900918-4

al. have shown that the MoS$_2$/rGO nanocomposite exhibited a reversible capacity of about 305 mAh g^{-1} at a current density of 100 mA g^{-1} after 50 cycles [61].

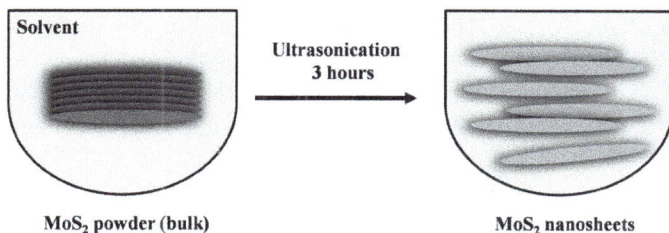

Figure 4. Solvent assisted liquid-phase exfoliation of MoS$_2$ nanosheets.

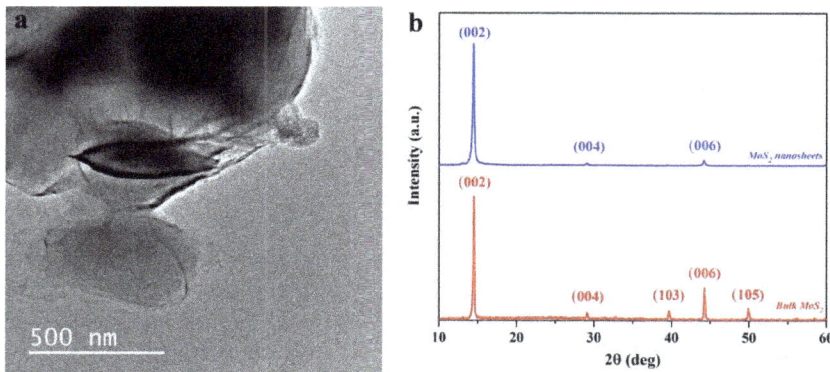

Figure 5. (a) TEM micrograph of the MoS$_2$ nanosheets prepared by liquid-phase exfoliation; (b) XRD pattern of the exfoliated MoS$_2$ nanosheets compared with that of the bulk MoS$_2$ powder.

7.5   Tungsten disulphide (WS$_2$)

Tungsten disulphide (WS$_2$) is a two-dimensional layered material like MoS$_2$. The WS$_2$ is synthesised in different nanostructure forms, such as nanoflowers [62], nanowires [63], nanotubes [64] and nanoflakes [65]. Cao *et al.* synthesised WS$_2$ of different morphology by following hydrothermal method [66]. The WS$_2$ nanoparticles, nanorods, nanosheets and nanofibers were synthesised using sodium tungastate and sulfourea as the source

Materials Research Forum LLC
https://doi.org/10.21741/9781644900918-4

material. But the surfactant (CATB/PEG) used in the synthesis process influenced the resultant nanostructures.

Feng *et al.* synthesised nanoflakes of WS_2 by following hydrothermal method at the temperature of 200 °C [65]. The produced WS_2 nanoflakes was used to prepared the anode material by mixing it with the carbon black and PTFE solution. The Li metal foil was used as the counter electrode and $LiPF_6$ was used as the electrolyte. The capacity of 1690 mAh g^{-1} was observed for the first cycle. But the passivation formed on the electrode surface due to the degradation of the electrolyte brought the capacity to the value of 680 mAh g^{-1}. This value was maintained for the remaining 20 cycles.

The mesoporous WS_2 structure was synthesised by vacuum assisted impregnation method [67]. The solution of phosphotungstic acid mixed with the mesoporous silica template was prepared. The dried mixture was undergone the sulphuration at 600 °C to form porous WS_2. The electrode prepared from this mesoporous WS_2 was tested. The initial capacity was reported as 1323 mAh g^{-1} at the current density of 0.1 A g^{-1}. The capacity was reduced to 805 mAh g^{-1} and it remained almost same for 100 cycles [67]. The material also showed the capacity of 503 mAh g^{-1} at the high current density of 10 A g^{-1}. The highly porous structure of WS_2 was the reason for this high capacity value that offered high surface area for electrolyte interaction, easy diffusion for ions and short pathways for the intercalation of lithium ions.

Du *et al.* synthesised oleylamine coated WS_2/graphene nanosheets by following modified colloidal synthesis method [68]. As the first step, a solution of WCl_6 in oleic acid was prepared. Then the solution was bubbled by N_2 followed by the addition of OLA into the solution. Then a small amount of CS_2 was added. After that this precursor solution was injected into a hot flask containing OLA. After injecting the precursor solution, the solution was rapidly cooled to room temperature. The obtained OLA coated WS_2 nanosheets was washed and dried. Then dispersion of OLA-WS_2 was sonicated followed by the addition of GO/ethanol to form the aggregation of OLA coated WS_2/GO nanosheets. The prepared composite was tested as the anode material in a coin-type lithium-ion battery. The initial capacity was reported as 542 mAh g^{-1} at the current density of 0.5 A g^{-1}. It retained the capacity value of 486 mAh g^{-1} for another 200 cycles. It even retained the capacity of 126 mAh g^{-1} at the current density of 10 A g^{-1}. In another investigation, Su *et al.* synthesised WS_2/graphene component by following hydrothermal process route [69]. It is used in sodium ion batteries and similar trend is expected in the lithium-ion batteries also. The WS_2/graphene nanocomposite anode owned a reversible specific capacity 540 mAh g^{-1} and a retention capability of 280 mAh g^{-1} after 500 cycles.

Wang *et al.* synthesised metallic (1T) WS_2 on semiconductor (2H) WS_2 by hydrothermal route [70]. The synthesised WS_2 (2H phase) was heat-treated in the presence of ammonia. This induced the conversion of 2H phase into 1T phase. The presence of metallic phase of WS_2 increased the conductivity of the material. The electrode material was prepared using 1T@2H WS_2 on the flexible carbon substrate. The electrochemical results reported the capacity value of 1130 mAh g^{-1} at the current density of 0.1 A g^{-1} for 200 cycles and the capacity value of 510 mAh g^{-1} at the current density of 2 A g^{-1} for 800 cycles [70].

7.6   Iron disulphide (FeS_2)

Iron disulphide (FeS_2) is widely used as the electrode material for non-rechargeable lithium-ion batteries. Iron disulphide has two phases, pyrite with the cubic crystal structure and marcasite with the orthorhombic crystal structure. Many researchers have reported that FeS_2, both pyrite and marcasite, as potential anode and cathode materials for low cost battery applications [71].

The FeS_2-pyrite nanorod clusters were synthesised by following low pressure hydrothermal route by our group [72]. Ferrous sulfate heptahydrate and sodium pyrosulfate were used as the source materials whereas sulphur was used to initiate the nucleation of the particles. The synthesis was carried out at 70 °C at normal pressure. Fig. 6a shows the SEM micrograph of the synthesised FeS_2. The FeS_2 was appeared as the clusters of the individual nanorods. The XRD patterns (Fig. 6b) of the FeS_2 shows that the peaks belongs to the pyrite-phase of the FeS_2.

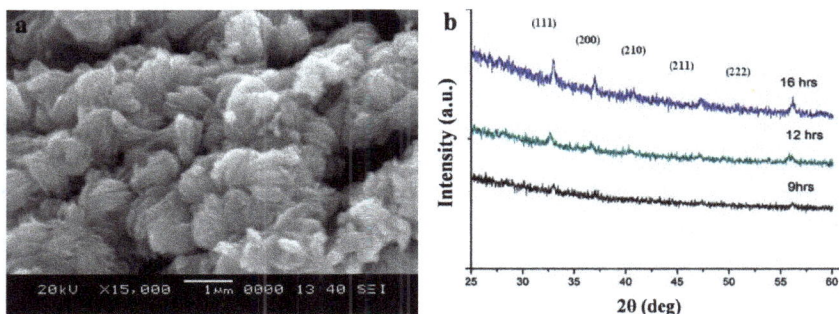

Figure 6. (a) SEM micrograph of the FeS$_2$ nanorods clusters (b) XRD pattern of the synthesised FeS$_2$-pyrite with different time durations.

There are many efforts to make nanocomposite using FeS_2 as one of the constituent. Biomass–carbon/FeS_2 nanocomposite was synthesized using carbonization and sulfuration and used as the cathode electrode material [73]. This composite had higher capacity to the tune of 850 mAh g^{-1} at 80 cycles. Xue et al. prepared the GO/FeS_2 composite by using the graphene oxide (GO), ferrous sulfate, sulphur and triethylene glycol as the precursor materials [74]. The pyrite, FeS_2 microspheres wrapped by the reduced graphene oxide sheets were obtained by heating the precursor material and followed by annealing. The prepared composite was used as the working electrode. The capacity calculated from the charge-discharge studies showed the excellent value of 970 mAh g^{-1} at the current density of 0.89 A g^{-1} for 300 cycles. It retained the capacity value of 380 mAh g^{-1} at the high current density of 8.9 A g^{-1} for 2000 cycles [74]. The graphene oxide with the FeS_2 microspheres offered higher conductivity, large surface area and improved interaction with the electrolyte. These were reasons behind this high and stable performance of the GO/FeS_2 electrode. The FeS_2 microspheres synthesised by the hydrothermal method was tested as the cathode material for the LIB [71]. The discharge capacity of the tested electrode material was reported as 371 mAh g^{-1} at 0.1 C and 239 mAh g^{-1} at 1 C for 50 cycles. To improve the performance of the electrode material, graphene was used along with the electrode material. The composite of rGO wrapped mesoporous FeS_2 particles were synthesised by facile solution reaction method [75]. The rGO provided the high surface area and conductive path to the electrode. This resulted in the high capacity value of 1720 mAh g^{-1} at the current density of 0.2 A g^{-1} for 700 cycles. The material retained the capacity of 340 mAh g^{-1} at the current density of 5 A g^{-1} at high temperature of 85 °C [75]. Fan et al. synthesised the marcasite FeS_2/carbon nanofibers composite by following hydrothermal method [76]. This metastable m-FeS_2 microparticles/CNF composite was tested as the anode material for the lithium-ion battery. It showed the high capacity value of 1086.9 mAh g^{-1} at the current density of 0.1 A g^{-1} and 782 mAh g^{-1} at the current density of 10 A g^{-1}. The cyclic stability test showed that the material retained the capacity of 575 mAh g^{-1} at 5 A g^{-1} after 1000 cycles. The main reasons for this good performance were, 1) the reduced electron/ion transfer pathways; (2) prevention of particle aggregation by the CNF; 3) structural stability provided by the CNF [76].

7.7 Tin sulphides

SnS_2 has layered structure. It consists of a layer of tin atoms located between two layers of hexagonally close packed sulphur atoms. The large interlayer spacing and the nanostructures of SnS_2 allows insertion and removal of the lithium ions. It can easily accommodate the volume changes associated with the cycling of lithium ions [77]. The SnS_2 has the capability of working at low voltage with high specific capacity. This feature has drawn the attention of many researchers to explore this material in the LIBs. There are many reports which aimed to enhance the electrochemical performance of the SnS_2 electrode materials. Wang et al. [78] have worked on the synthesis of SnS_2 nanosheets in N-methyl-2-pyrrolidore by following solvothermal method. The capacity exhibited by the SnS_2 nanosheets is more than 1200 mAh g^{-1} and better short-time cycling stability. The carbon coated SnS_2 also has shown the large capacity of ~600 mAh g^{-1} and decent cycling stability. Chen et al. [79] prepared the SnS_2/rGO nanocomposites by reflux condensation and hydrothermal methods. The capacity of SnS_2/rGO electrode is 514 mAh g^{-1} at 1.2 A g^{-1} after 300 cycles. The discharge capacity and cycling performance of the SnS_2/rGO is better compared to the SnS_2/C electrode. This is because of the introduction of rGO in the nanocomposites. Wang et al. [80] have synthesized the carbon coated SnS_2 by solid-state reaction method. The carbon coated SnS_2 nanocomposite shows the reversible capacity of 660 mAh g^{-1} at a current density of 50 mA g^{-1} and maintain up to 570 mAh g^{-1} for 100 cycles with a degradation rate of 0.14 %. The ex-situ characterization of SnS_2/carbon coated electrode gives better performance due to the stable morphology and structural integrity during charge-discharge cycles. Wang et al. [81] developed a light weight, high mechanical strength and high power/energy density flexible flower-like SnS_2 nanoplates decorated on the graphene nanosheets supported on the carbon cloth electrode. The initial capacity of the SnS_2/graphene is found to be 1987.4 mAh g^{-1} and specific capacity equal to 638.1 mAh g^{-1} after 150 cycles. The flower-like SnS_2 nanoplates decorated on the graphene nanosheets provide high surface area, excellent electrochemical performances and reduces electrode polarization. Liu et al. [82] demonstrated the synthesis of hexagonal sheet-like SnS_2/GO nanocomposites. The presence of small amount of GO (2.4 weight %) significantly improved the specific capacity, rate capability and efficient cycling performances. The initial charge and discharge capacity is 1057.9 mAh g^{-1} and 1449.7 mAh g^{-1}, respectively with Coulombic efficiency of 72.97 %. The discharge capacity still retains a good value of 509.5 mAh g^{-1} even after 100 cycles. The capacity retains a value of 589.0 mAh g^{-1} at 50 mA g^{-1} current.

Wang et al. synthesised novel ultra long SnS_2 nanobelts by following solvothermal method [83]. CS_2, dodecanethiol and tin(IV) chloride pentahydrate were used as the precursor materials to synthesis SnS_2 nanobelts. The nanobelts of several hundred micrometer length was synthesised. A coin cell battery was fabricated using the SnS_2 nanobelts as the working electrode Li foil as counter electrode and $LiFP_6$ as the electrolyte. The charge-discharge study revealed the initial charge capacity as 640 mAh g^{-1} and a steady capacity value of 560 mAh g^{-1} at current density of 0.1 C for 50 cycles [83].

Wang *et al.* prepared the SnS_2 nanosheets assembly to be used in the lithium-ion battery application [78]. The solvothermal method synthesised SnS_2 was treated with the dopamine hydrochloride and tris(hydroxymethyl) to produce carbon coated SnS_2 nanosheets. A R2032 type coin cell was fabricated by using the c-SnS_2 nanosheets as the working electrode, Li foil and $LiFP_6$ as the counter electrode and the electrolyte, respectively. The carbon coating over the surface of SnS_2 helped to facilitate the charge transportation and mechanical support. The capacity of the c-SnS_2 nanosheets was reported as 1100 mAh g^{-1} at the current density of 0.2 A g^{-1} for 50 cycles.

Zhang *et al.* synthesised three-dimensional honeycomb, like SnS_2 quantum dot/rGO composite for the lithium-ion battery application [84]. The polystyrene (PS) nanospheres were used as the template to synthesis 3D nanostructure. The tin (IV) chloride and rGO were mixed with the PS nanospheres and subjected to ultrasonication. The samples were heat treated to remove the polystyrene. Then the powder sample was mixed with thiourea and annealed to form 3D honeycomb-like SnS_2/rGO composite. The capacity value was calculated as 862 mAh g^{-1} at 0.1 A g^{-1} and 622 mAh g^{-1} at 0.5 A g^{-1} for 200 cycles.

7.8 Nickel Sulphides

Nickel sulphides, such as NiS, NiS_2 and Ni_3S_2 have been explored as the electrode material for the lithium-ion battery application. The nickel sulphides have good theoretical capacity of 550 mAh g^{-1} [85]. However, the cyclic performance of such material remains poor [86]. Mi *et al.* prepared NiS/Ni_3S_2 flower-like microspheres, nanoplates, prism-like microrod and chain-like nanostructures by using solvothermal method [85]. In brief, the Ni foam, ethylene glycol, anhydrous ethylenediamine and sulphur powder were used in this synthesis process. The volume ratio of ethylene glycol and anhydrous ethylenediamine was varied to produce different morphologies. The calculated discharge capacity of microspheres, nanoplates and 1D prism-like microrods & chain-like nanostructures was shown as 353.1, 445.5 and 550 mAh g^{-1} respectively at the current density of 40 mA g^{-1}. The shorter path for electron/lithium ion transportation and larger interface connection between the electrode and the electrolyte were the reason for the high discharge capacity value of the 1D nanostructures. To improve the performance, the nickel sulphides with various morphologies were synthesised to suppress the volume change during the charge-discharge cycle. Xiao *et al.* prepared nickel sulphide (NiS/Ni_3S_2) particles and interlinked nanosheets by in situ microwave synthesis [87].

A porous 3D nickel sulphide on Ni foam was synthesised by facile solution and melt–diffusion process [88]. The prepared material was tested as a potential cathode material for the LIBs. The discharge-charge study showed the initial discharge capacity of 1326 mAh g^{-1} and 715 mAh g^{-1} after 100 cycles at the current density of 0.2 C (118 mA g^{-1}).

Tao *et al.* synthesised NiS nanorods/N-doped graphene composite by following hydrothermal method. The prepared composite was tested as an anode for the LIBs [89]. The discharge-charge study revealed the charge capacity of the composite material as 687 mAh g^{-1}, whereas the bare NiS showed the charge capacity of 575 mAh g^{-1} at the current density of 50 mA g^{-1}. But the capacity of bare NiS reduced drastically to 100 mAh g^{-1} after 50 cycles. This is because of the particle aggregation and electrode pulverization during cyclic process. On the other hand, the composite NiS/N-doped graphene showed good capacity retention of 425 mAh g^{-1} after 50 cycles. The reason behind this stable capacity was the addition of N-doped graphene. It acted as the buffer during cyclic process to prevent the volume changes. The interface resistance was also reduced due to the high electrical conductivity of the graphene.

Dong *et al.* synthesised NiS nanoparticles embedded in the mesoporous N and S doped tubular carbon structures by a combination of nickel salt impregnation and calcinations process [90]. The synthesised NiS/NSC exhibited high surface area to the level of 200.8 m^{2}g^{-1}. The charge capacity of the prepared material was 715.9mAh g^{-1} at the current density of 0.1 A g^{-1} after 200 cycles with almost 100% Coulomb efficiency. The reported capacity value was higher than its theoretical capacity of 590 mAh g^{-1}. The NiS particles embedded on the carbon may promote the formation of polymer-like layer on their surface that can contribute further capacity value during the long cyclic process. The material showed the excellent capacity values of 446.1 mAh g^{-1} and 411.1 mAh g^{-1} even at the high current density of 2 A g^{-1} and 5 A g^{-1}, respectively.

8. Synthesis techniques for metal sulphides

8.1 Solid state method

This method utilizes two metals and sulfur. The two metals and sulfur are mixed together and heated to high temperature under vacuum or inert atmosphere condition [91]. The method has several disadvantages, like need of high temperatures (> 500 °C) and long reaction time [92]. In addition, this method offers less control over the size, morphology and structure for the synthesized nanosufides. The materials prepared with this method are chromium chevrel phase sulphide (Cr$_x$Mo$_6$S$_{8-y}$), MnS, Cr$_2$S$_3$, etc. In contrast, there are several low-temperature and solution-based methods which are more popular and promising.

8.2   The hydro/solvothermal method

This method is simple and widely used for the preparation of metal sulphides for LIBs. In this method, two metal salts and a source of sulfur are dissolved either in water or organic solvent or mixture of organic solvent and water. Then the solution is heated to a designed temperature for a particular time duration. The reaction for the hydro/solvothermal method is carried out in a sealed teflon-lined stainless steel autoclave where the autogenous pressure is generated inside the vessel. The pictorial representation of hydrothermal synthesis approach is as shown in Fig. 7. These conditions are sufficient for the nano/micro metal sulphide materials to precipitate and form a particular structure. The hydro/solvothermal route is effective and has good control over shape, size and morphology of the material. When metal foams are introduced as the source material, it is possible to construct metal-containing sulphide nanostructures on the source metal foam under hydro/solvothermal conditions. This method is environmental friendly and cost effective. The hydrothermally synthesized metal sulphides have high crystallinity with small crystallite size and highly uniform particles. The hydrothermal synthesis of various MoS_2-carbon based nanocomposites morphologies like nanosheets, nanoflakes, nanoplates, nanoflowers, microspheres were reported [22].

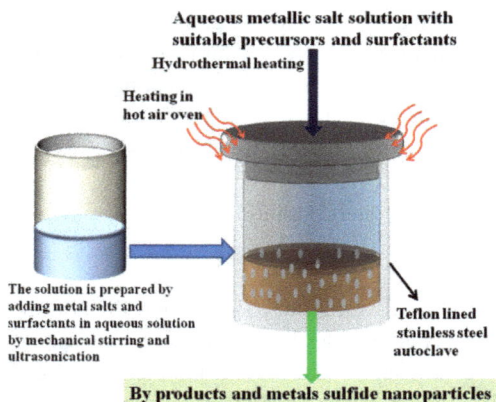

Figure 7. Pictorial representation of hydrothermal method for synthesizing metal sulphide nanoparticles.

There are many reports which followed hydrothermal synthesis method for metal sulphides and composites of metal sulphides nanostructures, due to its inherent advantages. Nanorod like CuS and Cu_2S are fabricated by the hydrothermal method without using any surfactant and template [22,28]. The nanocomposite of CoS_2/carbon is prepared by a freeze-drying technique and hydrothermal approach with the CoS_2 nanoparticles (~20nm) arranged on a spongy carbon matrix [93]. Wang *et al.* [81] synthesized flower like SnS_2 decorated on the graphene sheets by following hydrothermal method. The MoS_2 nanoparticles with different morphologies are synthesized by hydrothermal approach [94,95]. The Co_9S_8 nano hexagonal flake crystals are synthesized by hydrothermal route using cobalt sulphate hydrate, hydrazine hydrate solvent and sodium sulphate [96]. The average crystallite size of Co_9S_8 nanocrystals is 2.5 nm. Nickel sulphide nanoparticles (NiS and NiS_2) [97,98] is synthesized by hydrothermal conditions. Carbonaceous materials (like CNT, GO, graphene) combined with metal sulphide nanoparticles synthesized by hydrothermal method have shown the best performance compared to their individual counterpart. Graphene metal sulphide nanocomposites are also synthesized and explored in the applications of LIBs. The graphene present in the composite reduces the agglomeration of metal sulphide nanoparticles and also helps in the electrochemical activity enhancement [99].

8.3   Microwave-assisted hydrothermal synthesis

Microwave-assisted hydrothermal synthesis technique is simple, quick and energy efficient technique. It works on the basis of capacity to absorb microwaves by target metal salts and transforms it into heat. The mechanism of heating can be divided into two parts, namely ionic conduction and dipole movement. When the polar solvent (Ex: water, ethanol, etc.) is used as the liquid medium, the microwaves induces rotation of dipoles within the liquid. The rotation force causes the polar molecules to align and relax in the direction of oscillating electromagnetic radiations.

CuS nanospheres interconnected by the carbon nanotubes (CNTs) have been successfully synthesized via a facile one-step microwave-assisted method [100]. Xiao *et al.* [87] synthesized CuS, NiS/Ni_3S_2 and FeS_2 nanoparticles with different morphologies by microwave irradiation approach method. The morphology of the nanoparticles is dependent on the irradiation temperature. The Cu foil supported CuS particles, nanobuds and nanosheets is synthesized with irradiation time of 20 min at 25 °C, 80 °C and 140 °C, respectively. Conductive agent and binder material free, Cu foil supported CuS is directly used as electrode material for LIBs. The CuS nanosheets exhibited a very good initial reversible capacity of 588 mAh g^{-1} at 56 mA g^{-1} with a small capacity fading rate of 0.17% per cycle for 100 cycles. The rate capability of CuS is found to be very high with the value of 463 mAh g^{-1} at 2.8 mA g^{-1}. The electrochemical performance of CuS nanosheets is far better compared to the CuS nanoparticles due to the layered structure and easy diffusion of lithium ion and electron into the electrode. Similarly, NiS/Ni_3S_2 particles and interlinked nanosheets are synthesized by microwave irradiation approach with irradiation time of 20 min and temperature of 150 °C and 180 °C, respectively. The FeS_2 nanoparticles, interlinked nanosheets and nanosheets are synthesized at irradiation temperature of 120 °C, 150 °C and 180 °C, respectively. Co_9S_8 nanoparticles are synthesized with the same conditions keeping irradiation temperature of 180 °C. NiS nanoparticles are synthesized by cyclic microwave irradiation approach [101]. SnS_2 flower-like nanostructure is synthesized by microwave irradiation approach [102]. The SnS_2 nanostructure exhibited initial charge capacity of 412 mAh g^{-1} at current 1 A g^{-1}. Similarly, nanocomposites of MoS_2/WS_2-rGO and graphene/CuS are also synthesized by microwave irradiation approach [61,103,104].

8.4   Spraying-related methods

The spray-related methods are simple approach to synthesize mainly spherical nanoparticles. NiS_2, Fe_3S_4, CuS, ZnS [105–107] nanoparticles are synthesized by following spray related approach. The spray solution is prepared by mixing oleyamine and precursors along with the suitable metal sulphide salts. The solution is sprayed on the substrate (like SiO_2, etc) maintained at either preheated or room temperature. The substrate can be post treated for tune the morphology of the nanoparticles.

9.   Summary

Lithium-ion batteries (LIBs) for 21st century applications demand high capacity and high capacity retentions, lower weight and cost, higher voltage availability, etc. These requirements force the research community for developing alternative battery materials. Metal sulphides due to their large capacitive enhancement compared to conventional materials, could be one of the modern electrode materials for the LIBs. But their applicability is restricted by the limitations of poor cycle stability and rate performance. This paper comprehensively reviews the research activity in the domain of metal sulphides as the new material for the LIBs, especially from the viewpoint of demands in 21st century. Salient features in the field of research highlights and general processing methods are reviewed.

References

[1]   C. Liu, Z.G. Neale, G. Cao, Understanding electrochemical potentials of cathode materials in rechargeable batteries, Mater. Today. 19 (2016) 109–123.

https://doi.org/10.1016/j.mattod.2015.10.009.

[2]   J.R. Miller, P. Simon, Materials science: Electrochemical capacitors for energy management, Science 321 (2008) 651–652. https://doi.org/10.1126/science.1158736.

[3]   M. Li, J. Lu, Z. Chen, K. Amine, 30 Years of Lithium-ion batteries, Adv. Mater. 30 30 (2018) 1800561. https://doi.org/10.1002/adma.201800561.

[4]   J.M. Tarascon, M. Armand, Issues and challenges facing rechargeable lithium batteries, Nature 414 (2001) 359–67. https://doi.org/10.1038/35104644.

[5]   M.S. Balogun, Y. Luo, W. Qiu, P. Liu, Y. Tong, A review of carbon materials and their composites with alloy metals for sodium ion battery anodes, Carbon N. Y. 98 (2016) 162–178. https://doi.org/10.1016/j.carbon.2015.09.091.

[6]   F. Zheng, M. Kotobuki, S. Song, M.O. Lai, L. Lu, Review on solid electrolytes for all-solid-state lithium-ion batteries, J. Power Sources 389 (2018) 198–213. https://doi.org/10.1016/j.jpowsour.2018.04.022.

[7]   X. Xu, W. Liu, Y. Kim, J. Cho, Nanostructured transition metal sulfides for lithium ion batteries: Progress and challenges, Nano Today 9 (2014) 604–630. https://doi.org/10.1016/J.NANTOD.2014.09.005.

[8]   J. Wang, Y. Li, X. Sun, Challenges and opportunities of nanostructured materials for aprotic rechargeable lithium-air batteries, Nano Energy 2 (2013) 443–467. https://doi.org/10.1016/j.nanoen.2012.11.014.

[9]   Z. Ma, X. Yuan, L. Li, Z.F. Ma, D.P. Wilkinson, L. Zhang, J. Zhang, A review of cathode materials and structures for rechargeable lithium-air batteries, Energy Environ. Sci. 8 (2015) 2144–2198. https://doi.org/10.1039/c5ee00838g.

[10]   M.R. Palacín, Recent advances in rechargeable battery materials: A chemist's perspective, Chem. Soc. Rev. 38 (2009) 2565–2575. https://doi.org/10.1039/b820555h.

[11]   N. Alias, A.A. Mohamad, Advances of aqueous rechargeable lithium-ion battery: A review, J. Power Sources 274 (2015) 237–251. https://doi.org/10.1016/j.jpowsour.2014.10.009.

[12]   P. Simon, Y. Gogotsi, Materials for electrochemical capacitors, Nat. Mater. 7 (2008) 845–854. https://doi.org/10.1038/nmat2297.

[13]   J. Zhu, R. Duan, S. Zhang, N. Jiang, Y. Zhang, J. Zhu, The application of graphene in lithium ion battery electrode materials, J. Korean Phys. Soc. 3 (2014) 1–8. https://doi.org/10.1186/2193-1801-3-585.

[14]   Y. Wang, H. Li, P. He, E. Hosono, H. Zhou, Nano active materials for lithium-ion batteries, Nanoscale 2 (2010) 1294–1305. https://doi.org/10.1039/c0nr00068j.

[15]   B. Xu, D. Qian, Z. Wang, Y.S. Meng, Recent progress in cathode materials research for advanced lithium ion batteries, Mater. Sci. Eng. R Reports 73 (2012) 51–65. https://doi.org/10.1016/j.mser.2012.05.003.

[16]   N. Zhou, E. Uchaker, H.Y. Wang, M. Zhang, S.Q. Liu, Y.N. Liu, X. Wu, G. Cao, H. Li, Additive-free solvothermal synthesis of hierarchical flower-like $LiFePO_4/C$ mesocrystal and its electrochemical performance, RSC Adv. 3 (2013) 19366–19374. https://doi.org/10.1039/c3ra42855a.

[17]   F. Brochu, A. Guerfi, J. Trottier, M. Kopeć, A. Mauger, H. Groult, C.M. Julien, K. Zaghib, Structure and electrochemistry of scaling nano $C-LiFePO_4$ synthesized by hydrothermal route: Complexing agent effect, J. Power Sources 214 (2012) 1–6. https://doi.org/10.1016/j.jpowsour.2012.03.092.

[18]   Z.J. Zhang, P. Ramadass, Lithium-Ion Battery Systems and Technology, in: Batter. Sustain., 2012: pp. 319–357. https://doi.org/10.1007/978-1-4614-5791-6_10.

[19]   H. Zhang, X. Yu, P. V. Braun, Three-dimensional bicontinuous ultrafast-charge and-discharge bulk battery electrodes, Nat. Nanotechnol. 6 (2011) 277–281. https://doi.org/10.1038/nnano.2011.38.

[20]   B. Kang, G. Ceder, Battery materials for ultrafast charging and discharging, Nature 458 (2009) 190–193. https://doi.org/10.1038/nature07853.

[21]   M. Chhowalla, H.S. Shin, G. Eda, L.-J. Li, K.P. Loh, H. Zhang, The chemistry of two-dimensional layered transition metal dichalcogenide nanosheets., Nat. Chem. 5 (2013) 263–75. https://doi.org/10.1038/nchem.1589.

[22]   J. Zhao, Y. Zhang, Y. Wang, H. Li, Y. Peng, The application of nanostructured transition metal sulfides as anodes for lithium ion batteries, J. Energy Chem. 27 (2018) 1536–1554. https://doi.org/10.1016/j.jechem.2018.01.009.

[23]   X. Xie, Z. Ao, D. Su, J. Zhang, G. Wang, MoS_2/graphene composite anodes with enhanced performance for sodium-ion batteries: The role of the two-dimensional heterointerface, Adv. Funct. Mater. 25 (2015) 1393–1403. https://doi.org/10.1002/adfm.201404078.

[24]   H. Nishide, K. Oyaizu, MATERIALS SCIENCE: Toward Flexible Batteries, Science 319 (2008) 737–738. https://doi.org/10.1126/science.1151831.

[25]   W. Pi, T. Mei, J. Li, J. Wang, J. Li, X. Wang, Durian-like NiS_2@rGO nanocomposites and their enhanced rate performance, Chem. Eng. J. 335 (2018) 275–281. https://doi.org/10.1016/j.cej.2017.10.142.

[26]   Q. Pan, J. Xie, T. Zhu, G. Cao, X. Zhao, S. Zhang, Reduced graphene oxide-induced recrystallization of NiS nanorods to nanosheets and the improved Na-storage properties, Inorg. Chem. 53 (2014) 3511–3518. https://doi.org/10.1021/ic402948s.

Materials Research Forum LLC
https://doi.org/10.21741/9781644900918-4

[27]   J.S. Nam, J.H. Lee, S.M. Hwang, Y.-J. Kim, New insights into the phase evolution in CuS during lithiation and delithiation processes, J. Mater. Chem. A. (2019). https://doi.org/10.1039/c9ta03008e.

[28]   X. Li, X. He, C. Shi, B. Liu, Y. Zhang, S. Wu, Z. Zhu, J. Zhao, Synthesis of one-dimensional copper sulfide nanorods as high-performance anode in lithium ion batteries, ChemSusChem 7 (2014) 3328–3333. https://doi.org/10.1002/cssc.201402862.

[29]   B. Jache, B. Mogwitz, F. Klein, P. Adelhelm, Copper sulfides for rechargeable lithium batteries: Linking cycling stability to electrolyte composition, J. Power Sources. 247 (2014) 703–711. https://doi.org/10.1016/j.jpowsour.2013.08.136.

[30]   A. Débart, L. Dupont, R. Patrice, J.M. Tarascon, Reactivity of transition metal (Co, Ni, Cu) sulphides versus lithium: The intriguing case of the copper sulphide, Solid State Sci. 8 (2006) 640–651 https://doi.org/10.1016/J.SOLIDSTATESCIENCES.2006.01.013.

[31]   M.T. McDowell, Z. Lu, K.J. Koski, J.H. Yu, G. Zheng, Y. Cui, In situ observation of divergent phase transformations in individual sulfide nanocrystals, Nano Lett. 15 (2015) 1264–1271. https://doi.org/10.1021/nl504436m.

[32]   X. Wang, Y. Wang, X. Li, B. Liu, J. Zhao, A facile synthesis of copper sulfides composite with lithium-storage properties, J. Power Sources 281 (2015) 185–191. https://doi.org/10.1016/j.jpowsour.2015.01.172.

[33]   A. Débart, L. Dupont, P. Poizot, J.-B. Leriche, J.M. Tarascon, A transmission electron microscopy study of the reactivity mechanism of tailor-made CuO particles toward Lithium, J. Electrochem. Soc. 148 (2002) A1266. https://doi.org/10.1149/1.1409971.

[34]   J.Y. Park, S.J. Kim, K. Yim, K.S. Dae, Y. Lee, K.P. Dao, J.S. Park, H.B. Jeong, J.H. Chang, H.K. Seo, C.W. Ahn, J.M. Yuk, Pulverization-tolerance and capacity recovery of copper sulfide for high-performance sodium storage, Adv. Sci. 6 (2019) 1900264. https://doi.org/10.1002/advs.201900264.

[35]   J. Zhao, Y. Zhang, Y. Wang, H. Li, Y. Peng, The application of nanostructured transition metal sulfides as anodes for lithium ion batteries, J. Energy Chem. 27 (2018) 1536–1554. https://doi.org/10.1016/J.JECHEM.2018.01.009.

[36]   J. Zhang, A. Yu, Nanostructured transition metal oxides as advanced anodes for lithium-ion batteries, Sci. Bull. 60 (2015) 823–838. https://doi.org/10.1007/s11434-015-0771-6.

[37]   C.H. Lai, K.W. Huang, J.H. Cheng, C.Y. Lee, B.J. Hwang, L.J. Chen, Direct growth of high-rate capability and high capacity copper sulfide nanowire array

cathodes for lithium-ion batteries, J. Mater. Chem. 20 (2010) 6638–6645. https://doi.org/10.1039/c0jm00434k.

[38]   C. Feng, L. Zhang, M. Yang, X. Song, H. Zhao, Z. Jia, K. Sun, G. Liu, One-pot synthesis of copper sulfide nanowires/reduced graphene oxide nanocomposites with excellent lithium-storage properties as anode materials for lithium-ion batteries, ACS Appl. Mater. Interfaces 7 (2015) 15726–15734. https://doi.org/10.1021/acsami.5b01285.

[39]   C. Feng, L. Zhang, Z. Wang, X. Song, K. Sun, F. Wu, G. Liu, Synthesis of copper sulfide nanowire bundles in a mixed solvent as a cathode material for lithium-ion batteries, J. Power Sources 269 (2014) 550–555. https://doi.org/10.1016/j.jpowsour.2014.07.006.

[40]   Z. Li, W. Li, H. Xue, W. Kang, X. Yang, M. Sun, Y. Tang, C.S. Lee, Facile fabrication and electrochemical properties of high-quality reduced graphene oxide/cobalt sulfide composite as anode material for lithium-ion batteries, RSC Adv. 4 (2014) 37180–37186. https://doi.org/10.1039/c4ra06067a.

[41]   X. Li, N. Fu, J. Zou, X. Zeng, Y. Chen, L. Zhou, W. Lu, H. Huang, Ultrafine cobalt sulfide nanoparticles encapsulated hierarchical N-doped carbon nanotubes for high-performance lithium storage, Electrochim. Acta 225 (2017) 137–142. https://doi.org/10.1016/j.electacta.2016.12.127.

[42]   A. Abdulla, A. Supervisor, X. Sun, Metal sulfides as anode for lithium ion and sodium ion battery recommended citation, 2017. https://ir.lib.uwo.ca/etd (accessed March 7, 2019).

[43]   X. Zhang, X. Jie Liu, G. Wang, H. Wang, Cobalt disulfide nanoparticles/graphene/carbon nanotubes aerogels with superior performance for lithium and sodium storage, J. Colloid Interface Sci. 505 (2017) 23–31. https://doi.org/10.1016/j.jcis.2017.05.028.

[44]   Y. Zhou, Y. Zhu, B. Xu, X. Zhang, K.A. Al-Ghanim, S. Mahboob, Cobalt sulfide confined in N-doped porous branched carbon nanotubes for lithium-ion batteries, Nano-Micro Lett. 11 (2019). https://doi.org/10.1007/s40820-019-0259-z.

[45]   J. Liu, C. Wu, D. Xiao, P. Kopold, L. Gu, P.A. Van Aken, J. Maier, Y. Yu, MOF-Derived Hollow Co9S8 Nanoparticles Embedded in Graphitic Carbon Nanocages with Superior Li-Ion Storage, Small 12 (2016) 2354–2364. https://doi.org/10.1002/smll.201503821.

[46]   W. Chen, T. Li, Q. Hu, C. Li, H. Guo, Hierarchical CoS_2@C hollow microspheres constructed by nanosheets with superior lithium storage, J. Power Sources 286 (2015)159-165. https://doi.org/10.1016/j.jpowsour.2015.03.154.

[47]   R. Jin, L. Yang, G. Li, G. Chen, Hierarchical worm-like CoS$_2$ composed of ultrathin nanosheets as an anode material for lithium-ion batteries, J. Mater. Chem. A 3 (2015) 10677–10680. https://doi.org/10.1039/c5ta02646f.

[48]   Y. Hu, D. Ye, B. Luo, H. Hu, X. Zhu, S. Wang, L. Li, S. Peng, L. Wang, A binder-free and free-standing cobalt sulfide@carbon nanotube cathode material for aluminum-ion batteries, Adv. Mater. 30 (2018) 1703824. https://doi.org/10.1002/adma.201703824.

[49]   Y. Zhou, D. Yan, H. Xu, J. Feng, X. Jiang, J. Yue, J. Yang, Y. Qian, Hollow nanospheres of mesoporous Co$_9$S$_8$ as a high-capacity and long-life anode for advanced lithium ion batteries, Nano Energy 12 (2015) 528–537. https://doi.org/10.1016/j.nanoen.2015.01.019.

[50]   H. Li, F. Xie, W. Li, B.D. Fahlman, M. Chen, W. Li, Preparation and adsorption capacity of porous MoS 2 nanosheets, RSC Adv. 6 (2016) 105222–105230. https://doi.org/10.1039/c6ra22414h.

[51]   Y. Tian, X. Zhao, L. Shen, F. Meng, L. Tang, Y. Deng, Z. Wang, Synthesis of amorphous MoS2 nanospheres by hydrothermal reaction, Mater. Lett. 60 (2006) 527–529. https://doi.org/10.1016/j.matlet.2005.09.029.

[52]   S.K. Park, S.H. Yu, S. Woo, J. Ha, J. Shin, Y.E. Sung, Y. Piao, A facile and green strategy for the synthesis of MoS$_2$ nanospheres with excellent Li-ion storage properties, CrystEngComm 14 (2012) 8323–8325. https://doi.org/10.1039/c2ce26447a.

[53]   B. Guo, Y. Feng, X. Chen, B. Li, K. Yu, Preparation of yolk-shell MoS$_2$ nanospheres covered with carbon shell for excellent lithium-ion battery anodes, Appl. Surf. Sci. 434 (2018) 1021–1029. https://doi.org/10.1016/j.apsusc.2017.11.018.

[54]   N. Chaudhary, M. Khanuja, Abid, S.S. Islam, Hydrothermal synthesis of MoS2nanosheets for multiple wavelength optical sensing applications, Sensors Actuators A Phys. 277 (2018) 190–198. https://doi.org/10.1016/j.sna.2018.05.008.

[55]   G. Yang, Y. Gu, F. Yan, J. Wang, J. Xue, X. Zhang, N. Lu, G. Chen, Chemical vapor deposition growth of vertical MoS2 nanosheets on p-GaN nanorods for photodetector application, ACS Appl. Mater. Interfaces 11 (2019) 8453–8460. https://doi.org/10.1021/acsami.8b22344.

[56]   D. Kathiravan, B.R. Huang, A. Saravanan, A. Prasannan, P. Da Hong, Highly enhanced hydrogen sensing properties of sericin-induced exfoliated MoS$_2$ nanosheets at room temperature, Sensors Actuators B Chem. 279 (2019) 138–147. https://doi.org/10.1016/j.snb.2018.09.104.

[57]   C. Prabukumar, M.M.J. Sadiq, D.K. Bhat, K.U. Bhat, Effect of solvent on the morphology of MoS$_2$ nanosheets prepared by ultrasonication-assisted exfoliation, in:

AIP Conf. Proc., 2018. https://doi.org/10.1063/1.5029660.

[58]   Y. Liu, L. Zhang, Y. Zhao, T. Shen, X. Yan, C. Yu, H. Wang, H. Zeng, Novel plasma-engineered MoS_2 nanosheets for superior lithium-ion batteries, J. Alloys Compd. 787 (2019) 996–1003. https://doi.org/10.1016/j.jallcom.2019.02.156.

[59]   T. Xiang, Q. Fang, H. Xie, C. Wu, C. Wang, Y. Zhou, D. Liu, S. Chen, A. Khalil, S. Tao, Q. Liu, L. Song, Vertical $1T-MoS_2$ nanosheets with expanded interlayer spacing edged on a graphene frame for high rate lithium-ion batteries, Nanoscale 9 (2017) 6975–6983. https://doi.org/10.1039/c7nr02003a.

[60]   G.S. Bang, K.W. Nam, J.Y. Kim, J. Shin, J.W. Choi, S. Choi, Effective liquid-phase exfoliation and sodium ion battery, ACS Appl. Mater. Interfaces7 (2014) 7084-7089. https://doi.org/10.1021/am4060222

[61]   W. Qin, T. Chen, L. Pan, L. Niu, B. Hu, D. Li, J. Li, Z. Sun, MoS_2-reduced graphene oxide composites via microwave assisted synthesis for sodium ion battery anode with improved capacity and cycling performance, Electrochim. Acta 153 (2015) 55–61. https://doi.org/10.1016/j.electacta.2014.11.034.

[62]   A. Prabakaran, F. Dillon, J. Melbourne, L. Jones, R.J. Nicholls, P. Holdway, J. Britton, A.A. Koos, A. Crossley, P.D. Nellist, N. Grobert, WS_2 2D nanosheets in 3D nanoflowers, Chem. Commun. 50 (2014) 12360–12362. https://doi.org/10.1039/c4cc04218b.

[63]   G.A. Asres, A. Dombovari, T. Sipola, R. Puskás, A. Kukovecz, Z. Kónya, A. Popov, J.F. Lin, G.S. Lorite, M. Mohl, G. Toth, A. Lloyd Spetz, K. Kordas, A novel WS_2 nanowire-nanoflake hybrid material synthesized from WO_3 nanowires in sulfur vapor, Sci. Rep. 6 (2016) 1–7. https://doi.org/10.1038/srep25610.

[64]   S.M. Ng, H.F. Wong, W.C. Wong, C.K. Tan, S.Y. Choi, C.L. Mak, G.J. Li, Q.C. Dong, C.W. Leung, WS_2 nanotube formation by sulphurization: Effect of precursor tungsten film thickness and stress, Mater. Chem. Phys. 181 (2016) 352–358. https://doi.org/10.1016/j.matchemphys.2016.06.069.

[65]   C. Feng, L. Huang, Z. Guo, H. Liu, Synthesis of tungsten disulfide (WS_2) nanoflakes for lithium ion battery application, Electrochem. Commun. 9 (2007) 119–122. https://doi.org/10.1016/j.elecom.2006.08.048.

[66]   S. Cao, T. Liu, S. Hussain, W. Zeng, X. Peng, F. Pan, Hydrothermal synthesis of variety low dimensional WS_2 nanostructures, Mater. Lett. 129 (2014) 205–208. https://doi.org/10.1016/j.matlet.2014.05.013.

[67]   H. Liu, D. Su, G. Wang, S.Z. Qiao, An ordered mesoporous WS_2 anode material with superior electrochemical performance for lithium ion batteries, J. Mater. Chem. 22 (2012) 17437–17440. https://doi.org/10.1039/c2jm33992g.

[68]   Y. Du, X. Zhu, L. Si, Y. Li, X. Zhou, J. Bao, Improving the anode performance of WS_2 through self-assembled double carbon coating, (2015) 1–8.

[69]   D. Su, S. Dou, G. Wang, WS_2@graphene nanocomposites as anode materials for Na-ion batteries with enhanced electrochemical performances, Chem. Commun. 50 (2014) 4192–4195. https://doi.org/10.1039/c4cc00840e.

[70]   T. Wang, C. Sun, M. Yang, L. Zhang, Y. Shao, Y. Wu, X. Hao, Enhanced reversible lithium ion storage in stable 1T@2H WS_2 nanosheet arrays anchored on carbon fiber, Electrochim. Acta 259 (2018) 1–8. https://doi.org/10.1016/j.electacta.2017.10.154.

[71]   Y. Tao, K. Rui, Z. Wen, Q. Wang, J. Jin, T. Zhang, T. Wu, FeS_2 microsphere as cathode material for rechargeable lithium batteries, Solid State Ionics 290 (2016) 47–52. https://doi.org/10.1016/j.ssi.2016.04.005.

[72]   Y. Cheng, J. Huang, J. Li, Z. Xu, L. Cao, H. Ouyang, J. Yan, H. Qi, SnO_2/super P nanocomposites as anode materials for Na-ion batteries with enhanced electrochemical performance, J. Alloys Compd. 658 (2016) 234–240. https://doi.org/10.1016/j.jallcom.2015.10.212.

[73]   X. Xu, Z. Meng, X. Zhu, S. Zhang, W.Q. Han, Biomass carbon composited FeS_2 as cathode materials for high-rate rechargeable lithium-ion battery, J. Power Sources. 380 (2018) 12–17. https://doi.org/10.1016/j.jpowsour.2018.01.057.

[74]   H. Xue, D.Y.W. Yu, J. Qing, X. Yang, J. Xu, Z. Li, M. Sun, W. Kang, Y. Tang, C.S. Lee, Pyrite FeS_2 microspheres wrapped by reduced graphene oxide as high-performance lithium-ion battery anodes, J. Mater. Chem. A. 3 (2015) 7945–7949. https://doi.org/10.1039/c5ta00983j.

[75]   Y. Du, S. Wu, M. Huang, X. Tian, Reduced graphene oxide-wrapped pyrite as anode materials for Li-ion batteries with enhanced long-term performance under harsh operational environments, Chem. Eng. J. 326 (2017) 257–264. https://doi.org/10.1016/j.cej.2017.05.111.

[76]   H.H. Fan, H.H. Li, K.C. Huang, C.Y. Fan, X.Y. Zhang, X.L. Wu, J.P. Zhang, Metastable Marcasite-FeS_2 as a New Anode Material for Lithium Ion Batteries: CNFs-Improved Lithiation/Delithiation Reversibility and Li-Storage Properties, ACS Appl. Mater. Interfaces. 9 (2017) 10708–10716. https://doi.org/10.1021/acsami.7b00578.

[77]   B. Qu, C. Ma, G. Ji, C. Xu, J. Xu, Y.S. Meng, T. Wang, J.Y. Lee, Layered SnS_2-reduced graphene oxide composite - A high-capacity, high-rate, and long-cycle life sodium-ion battery anode material, Adv. Mater. 26 (2014) 3854–3859. https://doi.org/10.1002/adma.201306314.

[78]   Y. Wang, J. Zhou, J. Wu, F. Chen, P. Li, N. Han, W. Huang, Y. Liu, H. Ye, F.

Zhao, Y. Li, Engineering SnS_2 nanosheet assemblies for enhanced electrochemical lithium and sodium ion storage, J. Mater. Chem. A. 5 (2017) 25618–25624. https://doi.org/10.1039/c7ta08056e.

[79] H. Chen, B. Zhang, J. Zhang, W. Yu, J. Zheng, Z. Ding, H. Li, L. Ming, D.A.M. Bengono, S. Chen, H. Tong, In-situ grown SnS_2 nanosheets on rGO as an advanced anode material for lithium and sodium ion batteries, Front. Chem. 6 (2018). https://doi.org/10.3389/fchem.2018.00629.

[80] J. Wang, C. Luo, J. Mao, Y. Zhu, X. Fan, T. Gao, A.C. Mignerey, C. Wang, Solid-State fabrication of SnS_2/C nanospheres for high-performance sodium ion battery anode, ACS Appl. Mater. Interfaces. 7 (2015) 11476–11481. https://doi.org/10.1021/acsami.5b02413.

[81] M. Wang, Y. Huang, Y. Zhu, X. Wu, N. Zhang, H. Zhang, Binder-free flower-like SnS_2 nanoplates decorated on the graphene as a flexible anode for high-performance lithium-ion batteries, J. Alloys Compd. 774 (2019) 601–609. https://doi.org/10.1016/j.jallcom.2018.09.378.

[82] Y. Liu, J. Zeng, J. Liu, X. Wang, C. Peng, R. Wang, R. Zhang, Hexagonal sheet-like tin disulfide@graphene oxide prepared by a novel two-step method as anode material for high-performance lithium-ion batteries, Mater. Lett. 237 (2019) 29–33. https://doi.org/10.1016/j.matlet.2018.11.053.

[83] J. Wang, J. Liu, H. Xu, S. Ji, J. Wang, Y. Zhou, P. Hodgson, Y. Li, Gram-scale and template-free synthesis of ultralong tin disulfide nanobelts and their lithium ion storage performances, J. Mater. Chem. A. 1 (2013) 1117–1122. https://doi.org/10.1039/c2ta00033d.

[84] Y. Zhang, Y. Guo, Y. Wang, T. Peng, Y. Lu, R. Luo, Y. Wang, X. Liu, J.K. Kim, Y. Luo, Rational Design of 3D Honeycomb-Like SnS 2 Quantum Dots/rGO Composites as High-Performance Anode Materials for Lithium/Sodium-Ion Batteries, Nanoscale Res. Lett. 13 (2018). https://doi.org/10.1186/s11671-018-2805-x.

[85] L. Mi, Q. Ding, W. Chen, L. Zhao, H. Hou, C. Liu, C. Shen, Z. Zheng, 3D porous nano/micro nickel sulfides with hierarchical structure: Controlled synthesis, structure characterization and electrochemical properties, Dalt. Trans. 42 (2013) 5724–5730. https://doi.org/10.1039/c3dt00017f.

[86] H. Ruan, Y. Li, H. Qiu, M. Wei, Synthesis of porous NiS thin films on Ni foam substrate via an electrodeposition route and its application in lithium-ion batteries, J. Alloys Compd. 588 (2014) 357–360. https://doi.org/10.1016/j.jallcom.2013.11.070.

[87] S. Xiao, X. Li, W. Sun, B. Guan, Y. Wang, General and facile synthesis of metal sulfide nanostructures: In situ microwave synthesis and application as binder-free cathode for Li-ion batteries, Chem. Eng. J. 306 (2016) 251–259.

https://doi.org/10.1016/J.CEJ.2016.05.068.

[88] J.J. Cheng, Y. Ou, J.T. Zhu, H.J. Song, Y. Pan, Nickel sulfide cathode for stable charge-discharge rates in lithium rechargeable battery, Mater. Chem. Phys. 231 (2019) 131–137. https://doi.org/10.1016/j.matchemphys.2019.04.024.

[89] H.C. Tao, X.L. Yang, L.L. Zhang, S.B. Ni, One-step synthesis of nickel sulfide/N-doped graphene composite as anode materials for lithium ion batteries, J. Electroanal. Chem. 739 (2015) 36–42. https://doi.org/10.1016/j.jelechem.2014.10.035.

[90] X. Dong, Z.P. Deng, L.H. Huo, X.F. Zhang, S. Gao, Large-scale synthesis of NiS@N and S co-doped carbon mesoporous tubule as high performance anode for lithium-ion battery, J. Alloys Compd. 788 (2019) 984–992. https://doi.org/10.1016/j.jallcom.2019.02.326.

[91] K. Suzuki, T. Iijima, M. Wakihara, Chromium Chevrel phase sulfide ($Cr_xMo_6S_{8-y}$) as the cathode with long cycle life in lithium rechargeable batteries, Solid State Ionics. 109 (1998) 311–320. https://doi.org/10.1016/S0167-2738(98)00074-5.

[92] X.Y. Yu, X.W. David Lou, Mixed metal sulfides for electrochemical energy storage and conversion, Adv. Energy Mater. 8 (2018) 1701592. https://doi.org/10.1002/aenm.201701592.

[93] Y. Zhang, N. Wang, C. Sun, Z. Lu, P. Xue, B. Tang, Z. Bai, S. Dou, 3D spongy CoS2 nanoparticles/carbon composite as high-performance anode material for lithium/sodium ion batteries, Chem. Eng. J. 332 (2018) 370–376. https://doi.org/10.1016/J.CEJ.2017.09.092.

[94] L. Luo, M. Shi, S. Zhao, W. Tan, X. Lin, H. Wang, F. Jiang, Hydrothermal synthesis of MoS 2 with controllable morphologies and its adsorption properties for bisphenol A, J. Saudi Chem. Soc. 23 (2019) 762-773. https://doi.org/10.1016/j.jscs.2019.01.005.

[95] X. Zheng, Y. Zhu, Y. Sun, Q. Jiao, Hydrothermal synthesis of MoS_2 with different morphology and its performance in thermal battery, J. Power Sources 395 (2018) 318–327. https://doi.org/10.1016/j.jpowsour.2018.05.092.

[96] Y.C. Chen, Y.G. Zhang, Hydrothermal synthesis of Co_9S_8 nanocrystalline aggregations and spectral study, Spectrosc. Spectr. Anal. 26 (2006) 1117–1119.

[97] B. Naresh, D. Punnoose, S.S. Rao, A. Subramanian, B. Raja Ramesh, H.J. Kim, Hydrothermal synthesis and pseudocapacitive properties of morphology-tuned nickel sulfide (NiS) nanostructures, New J. Chem. 42 (2018) 2733–2742. https://doi.org/10.1039/c7nj05054b.

[98] F. Soofivand, E. Esmaeili, M. Sabet, M. Salavati-Niasari, Simple synthesis, characterization and investigation of photocatalytic activity of NiS_2 nanoparticles using

new precursors by hydrothermal method, J. Mater. Sci. Mater. Electron. 29 (2018) 858–865. https://doi.org/10.1007/s10854-017-7981-4.

[99]   G. Pengbiao, Z. Shasha, T. Hao, Z. Rongmei, Z. Li, C. Shuai, X. Huaiguo, P. Huan, Transition metal sulfides based on graphene for electrochemical energy storage, Adv. Energy Mater. 8 (2018) 1703259. https://doi.org/10.1002/aenm.201703259.

[100] Y. Wang, Y. Zhang, H. Li, Y. Peng, J. Li, J. Wang, B.-J. Hwang, J. Zhao, Realizing high reversible capacity: 3D intertwined CNTs inherently conductive network for CuS as an anode for lithium ion batteries, Chem. Eng. J. 332 (2018) 49–56. https://doi.org/10.1016/J.CEJ.2017.09.070.

[101] M. Salavati-Niasari, G. Banaiean-Monfared, H. Emadi, M. Enhessari, Synthesis and characterization of nickel sulfide nanoparticles via cyclic microwave radiation, Comptes Rendus Chim. 16 (2013) 929–936. https://doi.org/10.1016/j.crci.2013.01.011.

[102] Y. Zou, Y. Wang, Microwave solvothermal synthesis of flower-like SnS2 and SnO2 nanostructures as high-rate anodes for lithium ion batteries, Chem. Eng. J. 229 (2013) 183–189. https://doi.org/10.1016/j.cej.2013.05.119.

[103] D.H. Youn, C. Jo, J.Y. Kim, J. Lee, J.S. Lee, Ultrafast synthesis of MoS$_2$ or WS$_2$-reduced graphene oxide composites via hybrid microwave annealing for anode materials of lithium ion batteries, J. Power Sources 295 (2015) 228–234. https://doi.org/10.1016/j.jpowsour.2015.07.013.

[104] H. Li, Y. Wang, J. Huang, Y. Zhang, J. Zhao, Microwave-assisted Synthesis of CuS/Graphene Composite for Enhanced Lithium Storage Properties, Electrochim. Acta 225 (2017) 443–451. https://doi.org/10.1016/j.electacta.2016.12.117.

[105] D. Mondal, G. Villemure, G. Li, C. Song, J. Zhang, R. Hui, J. Chen, C. Fairbridge, Synthesis, characterization and evaluation of unsupported porous NiS$_2$ sub-micrometer spheres as a potential hydrodesulfurization catalyst, Appl. Catal. A Gen. 450 (2013) 230–236. https://doi.org/10.1016/j.apcata.2012.10.030.

[106] J.H.L. Beal, P.G. Etchegoin, R.D. Tilley, Transition metal polysulfide complexes as single-source precursors for metal sulfide nanocrystals, J. Phys. Chem. C. 114 (2010) 3817–3821. https://doi.org/10.1021/jp910354q.

[107] B. Geng, X. Liu, J. Ma, Q. Du, A new nonhydrolytic single-precursor approach to surfactant-capped nanocrystals of transition metal sulfides, Mater. Sci. Eng. B Solid-State Mater. Adv. Technol. 145 (2007) 17–22. https://doi.org/10.1016/j.mseb.2007.09.065.

Chapter 5

Magnetic Nanomaterials for Lithium-ion Batteries

Mine Kurtay[1], Haydar Göksu[2]*, Husnu Gerengi[1], Hakan Burhan[3], Mohd Imran Ahamed[4], Fatih Şen[3]*

[1] Corrosion Research Laboratory, Department of Mechanical Engineering, Faculty of Engineering, Duzce University, 81620, Duzce, Turkey

[2] Kaynasli Vocational College, Duzce University, 81900 Duzce, Turkey

[3] Sen Research Group, Department of Biochemistry, Dumlupınar, University, 43100 Kütahya, Turkey

[4] Department of Chemistry, Faculty of Science, Aligarh Muslim University, Aligarh-202002, India

haydargoksu@duzce.edu.tr, fatih.sen@dpu.edu.tr

Abstract

Nowadays, the rapid rise in technological developments has caused new devices in many fields, from health to communication. Not only it is limited to these areas, but also the speed of technology in individual uses has made it very easy to access many portable devices. The fact that mobile devices can offer advanced functional services to their users continuously is because they have a safe, long-lasting, high energy density, rechargeability, and environmentally friendly energy source. The main applications of lithium-ion batteries (LIB) are laptops, mobile phones, small household apparatus, and each of us prefers these electronic devices due to their non-toxicity and energy density. Rechargeable batteries are widely used in energy storage systems with the invention of lead-acid batteries. Thanks to high energy/power density, long cycle life, low power loss, and loop stability, LIBs are widely used as portable batteries in electronic devices and electric vehicles. The LIBs are intended to be used not only in the electronic field but also in commercial vehicles, smart grids, and large-scale energy storage. However, the properties of LIBs need to be improved. Different and various materials have been used to improve LIB. Among the materials used, nanomaterials are found in carbon-based structures. The small size of the nanomaterials provides high surface area efficiency and provides electrode and electrolyte contact. In addition, it extends the energy, power, and service life of LIBs. In this chapter, the studies and researches carried out have been organized in order to strengthen the LIBs based on nanomaterials for longer lasting effects.

Keywords

Magnetic Nanomaterial, Ion Batteries, Lithium, Energy Storage

Contents

1. Introduction

Nowadays, it is known that there is a rapid technological consumption in all areas from transportation to communication, from health to defense and almost every individual has at least one of the portable electronic devices [1]. In addition, most of the electrical appliances we use at home with the developing technology can be used wirelessly. Portable electronic products need to have a high energy density, safe, long life, easy to maintain, rechargeable and environmentally friendly energy source in order to fulfill their developed functions in a long and effective manner. Lithium-ion batteries (LIB) are preferred in the designing of electronic devices such as laptops, mobile phones, small household appliances due to their high energy densities and non-toxic properties [2,3]. Moreover, it is also preferred by environmentally conscious designers and consumers due to low CO_2 emissions [4].

Among nanomaterials, magnetic nanoparticles have a great interest from the researchers because of their excellent magnetic properties [6]. The nanoparticles having a size below 100 nanometers (nm) have properties which are considered to be different and superior to the volumetric structured materials. The nanoparticle [NP] materials offer great advantages due to their unique dimensions and physicochemical properties [7]. In recent years, thanks to the studies made about nanomaterials, more robust products, better quality, cheaper, longer lasting, lighter, and smaller in size are obtained [8].

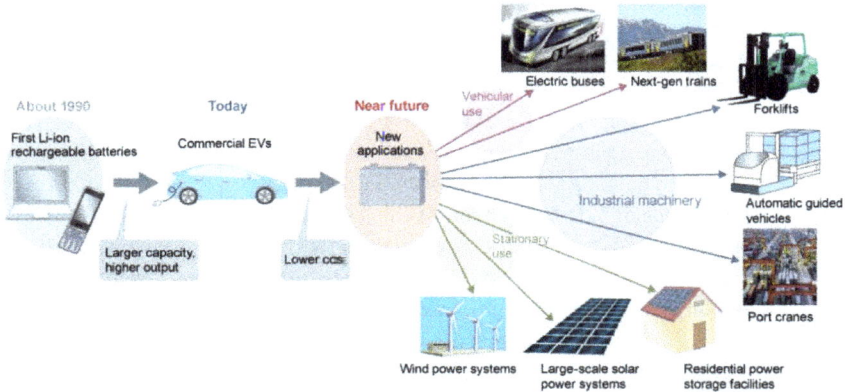

Fig. 1. Examples for the applications of LIBs [5].

Magnetic nanoparticles have been the focus of much research [9], including magnetic fluids recording [10], catalysis [11], biotechnology [12,13] and tissue specific targeting [14], material sciences, photo catalysis [15], electrochemical and bioeletrochemical sensing [16], microwave absorption [17], cation sensors [18], magnetic resonance imaging [MRI] [19], medical diagnosis [20], data storage [21, 22], nanofluids [23, 24], optical filters [25], defect sensor [26], magnetic cooling [27, 28], environmental remediation [29] and, as an electrode [30], for super capacitors and LIBs [31,32].

2.   History of LIBs

Lithium with a low atomic number has a higher electrode potential by comparison with lead and zinc in conventional batteries. The development of new anode, cathode, and anhydrous electrodes is needed for the improvement of high energy LIB systems. Gilbert N. Lewis first discovered the LIB in 1912. The first non-rechargeable batteries emerged in the early 1970s. The availability of rechargeable LIBs has been possible after nearly 20 years of work [33]. For many years, only nickel-cadmium batteries were available for use in portable devices (wireless communication). In 1990, nickel-metal hydride and LIBs with higher capacity were introduced. LIBs have become the main power-supply source for movable products owing to their low weight and high energy density. In 1991, Sony commercialized the first LIB, and today the LIB is the fastest-growing battery in the world. Lithium is the lightest metal which provides the highest energy content and has the highest electrochemical potential [34]. The main battery systems used today are primary manganese dioxide-zinc and rechargeable lead-acid. These battery systems with a deep-rooted history have advanced technical service life. As long as the request for high-performance battery systems remains, the LIB will replace existing systems. Nowadays, there are important developments in the energy storage systems of electric vehicles that come up more frequently in Tesla's automobiles. Samsung has opened a LIB factory in China by making investment plans in this area. Samsung SDI offers a wide range of applications of LIBs from hybrid vehicles to pure electric vehicles. In addition, the creation of two different application areas, namely the high voltage and low voltage system, is also important for meeting customer demands. Thus, energy needs are provided for both low power and high-power segments [35].

3. LIB Technology

Rechargeable batteries are called secondary batteries. They are the electrochemical cells that can be recharged after discharge. The advantages and disadvantages of LIBs are shown in Table 1 [36].

Table 1. Some positive and negative features of lithium-ion batteries.

Positive	Negative
• Closed cell, maintenance-free	• Relatively expensive
• Long cycle life	• Degradation at high temperatures
• Wide operating temperature range	• Requires protection circuitry for safety and to prevent overcharge and over discharge
• Long shelf life	
• Low self-discharge rate	
• Quick charging ability	• Not tolerant of overcharge and over-discharge
• High power discharge capacity	
• High energy potential	• Thermal runaway concerns
• High specific energy and energy density	
• No memory effect	

The most significant flaw of LIBs is that their life starts from the date of production. A notebook battery with a 100% charge level and a temperature of 25 °C in most cases loses 20% of its capacity each year. This loss of capacity starts from the date of production of the product and continues even if the battery is not used at all [37].

There are many metals which react with lithium for high capacity LIB applications. Today, graphite is used as anode material in most commercial LIBs. The disadvantages of the graphite used as anode material are low capacity and the occurrence of safety problems associated with the lithium ion into the structure of the anode [38-40]. Recent studies have focused on alternative anode materials with low cost, no security problems, high energy density, and long cycle count [41, 42]. In addition, lithium alloyed metals are remarkable as anode material for rechargeable lithium-ion batteries [39-40]. Among these alternative positive electrode materials, metals such as Al, Sn, Sb, and Si, which have a high specific capacity, and their metals capable of alloying with lithium are prominent [43-46]. The Sn-based anode materials have a discharge capacity of approximately 990 mAh/g [47, 48]. This discharge capacity is three times the discharge capacity of the graphite anode. However, during the charging and discharging processes, the anode material begins to pulverize decomposition at the end of a certain battery cycle, as the tin generates a volume expansion of approximately 400% [49-51]. Therefore, the production studies on tin-based anode materials are centered on intermetallic nanocomposites, when considering the active-inactive composites. Tin-cobalt-based materials are important in terms of active-inactive nanocomposite anode materials. Valvo et al. produced tin-cobalt anode material by the electro-spray method and argued that they improved the life of the anode material [52]. Chen et al. produced the Sn-Co based anode material with polymer pyrolysis method and developed anodized material which was more stable to use for longer period [53]. Li et al. capsulized the nanoscale $M_x Sn$ (M = Fe, Cr, and Ni) alloys into amorphous carbon nanotubes (ACNTs) and formed hybrid anode hetero-structures. As a result, they indicated that anodes with large volume change significantly reduced electrochemical degradation, and this improved LIB performance [54].

4. LIB working principle

In a rechargeable LIB, lithium ions move into an anode from a cathode during discharge in an aqueous or non-aqueous electrolyte and return to their poles during charging. Figure 2 shows the basic working principles of LIB.

Lithium-ion Batteries - Materials and Applications   Materials Research Forum LLC
Materials Research Foundations 80 (2020) 123-147   https://doi.org/10.21741/9781644900918-5

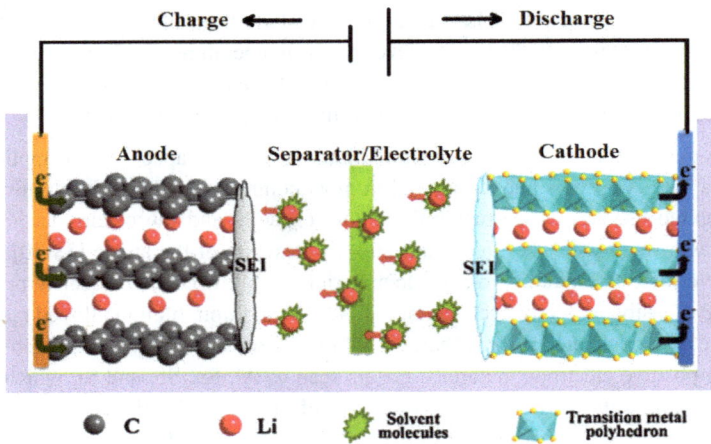

Fig. 2. Schematic of a typical LIB.

The half-reaction for charging takes place in the anode is given below:

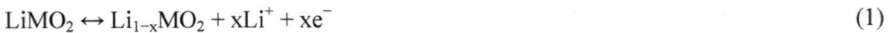

$$LiMO_2 \leftrightarrow Li_{1-x}MO_2 + xLi^+ + xe^- \tag{1}$$

And the half reaction for charging takes place in the cathode is given below:

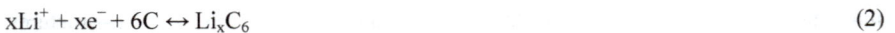

$$xLi^+ + xe^- + 6C \leftrightarrow Li_xC_6 \tag{2}$$

where x is a coefficient for reactions taking place the cathode and anode electrodes, and these reactions are in moles unites [55]. A LIB comprised of anode and cathode electrodes and operates as an energy storage device thanks to the conversion of the electrical energy to electrochemical energy. Both cathode and anode materials used in today's commercialized LIB systems are composite materials. The materials present on the anode and cathode electrodes have some specials sides made up a rigid structure for Li-ions. In the beginning, the battery system is "discharged" when all ions are on the negative electrode side. While charging, ions reach from the cathode to the anode by moving in the electrolyte. The transmission of electrons takes place from the cathode to anode thanks to an external collector, and that leads to a forming electrical current. The chemical energy potential of the anode is higher than the cathode that leads to an electrochemical energy storage form electrical energy. The electrochemical energy is converted to a form of electrical energy, upon the LIB is discharged. The LIB composed of two sections which are separated by a membrane made up of a microporous structure that prevents the electrolytes flowing between the electrodes. The electrolyte must be ionically conductive and electronically non-conducting, but the real properties of the electrolyte are much more complex. In present LIB technology, capacities and cell voltage are determined fundamentally by cathode material, which is a limiting factor for the transport rate of Li. Therefore, the development of negative electrode materials becomes extremely important and attracts a great deal of attention in the last decade [56].

5.   Nanomaterials

Nanomaterials are used to develop the speed properties of solid-state electrodes used in batteries. Nanomaterials should be elaborated in an electrode that sustains the diffusion length, besides ensuring electrical and mechanical contacts during the strain, carries out structural stresses, supporting structural stresses, and support a longer service life. Various methods have been used to prepare non-self-supporting nanometric materials [57-60]. With these approaches, nano-sized materials (rods, wires, spheres, etc.), which have improved electrochemical properties and need to be further processed in an electrode film have been obtained. New electrode configurations have emerged to maintain the benefits of electrochemistry on a nanoscale and to obtain high-speed capabilities [61]. The use of nanomaterial electrodes in LIBs enables the development of energy, power, and cycle life. Nanoparticles or nanopowder electrode materials, which are thin versions of micron-sized electrode powders in lithium-ion batteries, have been used for a long time. Carbon black is the best example of nanomaterials used in LIBs since time immemorial [57]. However, carbon black is the nanomaterial that does not store electrical energy and has only a passive conductivity enhancing properties to increase the power capacity. However, remarkable performance developments can be achieved for two reasons by designing the electrode's active energy storage component as a nanoparticle:

- Short diffusion lengths of lithium-ion moving from the particle core to the electrolyte,

- High electrode-electrolyte contact area caused by high surface areas of the particles [62].

In nanoscale particle size electrodes, the mechanical stresses are significantly reduced due to the volumetric expansion and contraction during charge and discharge. Studies have shown that particles should be smaller than a certain critical radius. Thus, instead of plastic deformation, elastic deformation, which allows the structure to be recovered without deterioration occurs [63]. Nanomaterials are used to maximizing the service life with minimum energy density in nanopowder form, nanostructured films with wire, rod, or column morphology [64-66].

Nanosize Li-ion storage electrode materials can provide various processing advantages during operation. Many materials made of lithium and other transition metals are used as lithium-ion cathode materials. Then, heat treatment is carried out under different conditions to obtain the desired phosphate or oxide composition. If the heat treatment is not applied correctly, the performance of the material may be reduced by causing a heterogeneous composition in the materials. During the process of forming nanomaterials by heat treatment, a homogeneous thermal profile should be maintained in all parts of the material. Thus, the production of non-heterogeneous compounds will be facilitated without the need for the application of intensive energy processing methods. It is hoped that nanomaterials will provide new generation high-speed battery formats for components such as sensors, RFID, and ultra-thin and flexible devices that can be used in medical applications [67, 68].

The major disadvantage of LIBs is the low power density due to high polarization taking place at the excess charge and discharge rate. The high polarization occurs due to weak electrical-thermal transmissions and low diffusion rate of Li-ion through the electrolyte and electrode interfaces. The development of high-energy rechargeable LIBs is important for the improvement of portable electronic devices and electric vehicles [69, 70]. The limited specific capacity of cutting-edge technology negative and positive electrode materials is the biggest barrier to high-energy LIBs [71, 72]. Developing new materials having high thermal-electronic conduction, and a larger surface and a low diffusion to use in electrodes is very important to overcome this problem. Nano-scale electrode materials have many advantages than conventional materials: (i) increased Li-ion input/output ratios owing to the short transport path; (ii) reversible Li insertion and removal without destroying the electrode material structure; (iii) increased contact area with electrolyte reduces intercalation-related volume changes; (iv) improved electron transport properties [73, 74]. Carbon nanotubes (CNTs) are used as anode or an additive for both positive and negative electrode materials to fulfill the above multifunctional requirements for LIBs [75-82].

Multi-walled carbon nanotubes (MWCNT) have been widely used owing to their excellent mechanical properties, electrical conductivity, high flexible structure, and high surface area. In addition, carbon nanotubes are rapidly replacing electrochemical energy storage systems, particularly in the development of LIB electrodes [83]. The conductivity network structure of the MWCNT with high conductivity helps the Sn materials to maintain their structure during the cycle. Moreover, the high conductivity and surface area of the MWCNT increase the capacity of the lithium to accommodate on the electrode surface and help to make the reaction faster between lithium and the active material [84].

6.   Nanomaterials in anode for LIBs

The use of nanomaterials in the anode of LIBs provides advantages such as low cost, high charge, and discharge rate, and energy density. Composite materials are used, or surface modification is applied to develop the stability of nanometer materials. For LIB systems, researches are being done to develop alternative anode materials that will provide higher energy density.

Positive electrode materials for LIBs should

(1) be able to store a larger amount of lithium reversibly,

(2) have electrochemical potential values of lithium extraction/insertion that are closely above lithium metal (0 V vs. Li^+/Li),

(3) have high ionic and electronic conductivities,

(4) have a small variation of the electrode structure with lithium storage,

(5) be abundant in non-toxic and earth,

(6) be economical for the manufacturer,

(7) be stable with the electrolyte in a wide range of electrochemical window.

Because of their high gravimetric capacities, lithium and its alloys formed with different metals are used as anode materials. Experimental studies have been carried out on lithium using as anode metallic or semi-metallic elements such as Ag, Al, Bi, Si, In, Au, Sn, Pb, Mg, Sb, Zn, Cd, Ge, As, Pt [85]. In addition, nano compounds have a great interest as positive electrode materials owing to their high theoretical capacity, abundant in nature, and environmentall friendliness [86]. Concerning anode materials, Si is considered the next generation anode because of its higher capacity than the traditional graphite anode, low operating potential, high abundance and environmentally friendly nature [87]. When Ge tested as a positive electrode for LIBs, it has indicated large capacity, high initial coulombic efficiency, and long cycle life owing to its self-healing mechanism and practical applicability [88]. Liu et al in their study, porous micro/nano lithium cerium oxide (Li_2CeO_3) with baseball morphology were prepared by precipitation and thermal annealing. The prepared Li_2CeO_3 (approximately 1μm diameter) was made from 5nm nanoparticles. As anode of lithium-ion batteries, the Li_2CeO_3 synthesized in the micro/nanocomposite structure provided a high discharge capacity of 582.7 mAhg^{-1} after 100 cycles at 0.5 mAcm^{-2} [89]. The 3D porous nano/micro $ZnFe_2O_4$ was prepared by sintering at 800°C in an air atmosphere after chemical precipitation. It is determined that $ZnFe_2O_4$ prepared as a positive electrode material for LIBs, displays high specific capacity, good cycle stability, and excellent speed capacity [90]. Silicone is an alternative as a high-performance anode material because it has a very high specific capacity and abundance. In the study, carbon-encapsulated nano-Si on graphite hybrid composite (nano-Si / G / C) was carefully designed and obtained with an easy and scalable process. The produced nano-Si / G / C anodes exhibit high cyclic stability with varying field mass loads ranging from 0.968 to 4.28 mg-cm^{-2}. Thanks to their superior properties, nano-Si / G / C composites have a high potential for use in lithium-ion batteries [91]. In another study, the mesoporous carbon and composite structures consisting of nano-sized Fe_3O_4 (10-20nm) particles were synthesized by a two-step method. The Fe_3O_4/C composite displayed a high specific capacity of 1498 mAhg^{-1} and a recycled capacity of 717 mAhg^{-1} after 100 cycles of 100 mAg^{-1} currents, and it achieved excellent cycle performance [92]. VBO_3 composite nanoparticles (VBO_3@G) on graphene plates were synthesized as positive electrode material for LIBs. Graphene layers helped to form nanoparticles and provide an excellent conductive 3D network. It was found that the composite exhibited a good cycle performance with an initial charging capacity of 779 mAhg^{-1} at 0.1 C and maintained at 610 mAhg^{-1} after 100 cycles [93]. Jia et al. reported the superior cycling stability and speed performance of the CuO electrode in LIBs. The CuO NL anodes provided a special capacity of 320 mAhg^{-1} after 800 cycles and exceptionally high-speed performance. This was direct result of the unique synergistic effects of hierarchical pores and vibrating nanostructures. This study supported the development of electrode materials for long-life energy storage devices [94]. Metal sulfides are considered as electrode materials in batteries because of their high theoretical capacity. Wang et al. proposed a different strategy for producing CoS and Co9S8 electrodes for advanced LIBs. As a result of the study, it was determined that porous cobalt sulfide/carbon composites, especially CoS/C/CNT (CoS-0.4C) and Co9S8/C/ CNT (Co9S8-0.8C), which were mounted as LIB anodes, exhibited superior lithium storage properties [95]. The nano-porous nano-silicon (npn-Si) anode for the LIB reduces the problems associated with pulverization and large-volume expansion during the charge-discharge cycle. The study shows that the Na salt of the carboxymethyl cellulose (Na-CMC) and alginate (Na-Alg) have good cycle stability and speed performance [96]. In another study, nickel silicate (Ni_2SiO_4, NSO) was synthesized as a high-capacity anode material for rechargeable LIBs. The results were evaluated as a promising positive electrode candidate for LIBs due to NSO nanosheets, excellent rate capability, and good loop stability [97]. The nanocomposite consisting of multi-walled carbon nanotubes modified from hematite nanoparticles has been tested as anode material of LIBs [98]. For LIBs, it is extremely important to develop graphite alternative anode materials at low temperature (below 0 ° C) for safety and capacity [99]. A multi-phase composite consisting of TiO_2/TiN/graphene was designed to improve the ionic and electronic conductivities and improve the electrochemical performance at low temperature. The nanocomposites produced showed excellent rate and superior electrochemical performance in low temperature (0, -10 and -20 °C) lithium-ion batteries. Therefore, the prepared nanocomposites have the potential for use in LIBs [100].

7. Nanomaterials in cathode for LIBs

The negative electrode materials used in LIBs are generally spinel or layered crystal structure. Cathodes determine the capacity and cycle life of LIBs. In LIBs, $LiMO_2$ type structures (M: Co, Ni, Mn, V) are used as cathode active substances [101-106]. These structures are suitable for Li^+ during charging and discharging [107]. The cathode active substances should be readily prepared and have high operating voltage, high binder efficiency, high capacity, chemical stability, and long cycle life. In addition, the operating voltage should not change much during discharge [108]. In recent studies, the emphasis has been focused on cathodes with cheaper, environmentally friendly, high capacity, and cycle life instead of commercially available cathodes. The reaction surface area and the capacity value increase due to the small particle size in the active cathode material.

Ni et al. synthesized bismuth oxyfluoride impregnated CMK-3 nanocomposite. The bismuth oxyfluoride nanoparticles produced have shown cycle stability and high capacity when used as the cathode of LIBs [109]. Graphene/a-MnO_2 nanocomposite cathodes were produced, and the charge-discharge properties of $Li-O_2$ cells with composite cathodes were investigated by Çetinkaya et al. This study showed that graphene/α MnO_2 nanocomposite cathodes on nickel foam can be used as cathodes for $Li-O_2$ batteries [110]. The structure and electrochemical activity of the electrocatalysts are important for lithium-oxygen batteries. Mn and Ru oxides were synthesized on multi-walled carbon nanotubes (RuO_2-MnO_2/MWCNTs) and used as a cathode [111]. In another study, a nanocomposite of V_2O_5 / acetylene black (V_2O_5/C) was synthesized for use as negative electrode material in LIBs. Stable electrochemical performance of V_2O_5/C nanocomposite showed that it can be used as a cathode material for LIBs [112]. Te nanotubes developed on solvent-free, and carbon substrates provide excellent rate capability due to their electrochemical performance and synergistic effect. The results showed that the Te / CFC can be used for energy storage and portable electronic devices in rechargeable Li-Te batteries [113]. Poly-(1,4-anthraquinone)/carbon nanotube

(P14AQ/CNT) nanocomposite has been developed to improve the rate performance of LIBs. Results showed that The P14AQ/CNT composite used as a cathode material for LIB had excellent rate performance and cycle stability [114]. Cubic spinel lithium manganese oxide ($LiMn_2O_4$) is a hopeful cathode material for large-scale LIBs. Ultra-thin 1D $LiMn_2O_4$ nanorods can be used as a promising cathode nanocomposite because they reduce Li-ion diffusion with their morphology and crystal structure [115]. Multi-walled carbon nanotube (MWCNT) -LiV_3O_8 nanocomposites were produced by the sol-gel method with no additive. The electrochemical performance of MWCNT - LiV_3O_8 nanocomposites was significantly improved [116]. As a summary, many types of nano-based materials have been used for many types of applications [117-135] and battery systems have also very important application areas for this type of nanomaterial.

Conclusions

With the development of the industry, the cost-efficient and efficient storage of energy from renewable resources have become an important issue. Rechargeable batteries are widely used in energy storage systems with the invention of lead-acid batteries. Thanks to its high energy/power density, long cycle life, low power loss and cycle stability, LIBs are widely used as portable batteries in electronic devices and electric vehicles. However, the development of electric vehicles is limited since the energy density of LIBs is not high enough. Despite their remarkable successes in commercial applications, there is a need to improve the properties of LIBs for large-scale energy storage in hybrid vehicles and smart grids. Therefore, it is focused on new generation batteries with high energy density in studies.

References

[1] M.,Yoshio, R.J. Brodd, A. Kozawa, Lithium-Ion Batteries:Science and Technologies, Springer Science and Business Media, Newyork, USA, (2009) 1-7. https://doi.org/10.1007/978-0-387-34445-4

[2] N. Bahaloo-Horeh, S. Mohammad Mousavi, M. Baniasadi, Use of adapted metal tolerant Aspergillus niger to enhance bioleaching efficiency of valuable metals from spent lithium-ion mobile phone batteries, Journal of Cleaner Production, 197 (2018) 1546-1557. https://doi.org/10.1016/j.jclepro.2018.06.299

[3] G. Zubi, R. Dufo-López, M. Carvalho, G. Pasaoglu, The lithium-ion battery: State of the art and future perspectives, Renewable and Sustainable Energy Reviews, 89 (2018) 292-308. https://doi.org/10.1016/j.rser.2018.03.002

[4] S.A. Hackney, R.V. Kumar, High Energy Density Lithium Batteries, Wiley-VCH Verlag GmbH, Weinheim, (2010) 70-73.

[5] H.C. Çoban, Metal oxide (SnO2) Modified LiNi0.8Co0.2O2 cathode material for lithium ion batteries, istanbul technical university, Department of Nano Science and Nano Engineering Nano Science and Nano Engineering Programme, M.Sc. Thesis, May 2014.

[6] A. Akbarzadeh, M. Samiei, S. Davaran, Magnetic nanoparticles: preparation, physical properties, and applications in biomedicine, Nanoscale Research Letters, 7(144) (2012) 1-13. https://doi.org/10.1186/1556-276X-7-144

[7] I. Bychko, E.Y. Kalishin, P. Strizhak, Effect of the size of Fe@ Fe3O4 nanoparticles deposited on carbon nanotubes on their oxidation-reduction characteristic, Theoretical and Experimental Chemistry, 47(4) (2011) 219-224. https://doi.org/10.1007/s11237-011-9207-9

[8] T. Baykara, Nanoteknoloji ve Nano-Yapılı Malzemeler, Nanoteknoloji ve Nanomalzeme Süreçleri, Tübitak Mam, Gebze, 2 (2010) 17-46.

[9] K. Khan, S. Rehman, H.U. Rahman, Q. Khan, Synthesis and application of magnetic nanoparticles, Nanomagnetism, (2014) 135-159.

[10] S. Singamaneni, V.N. Bliznyuk, C. Binek, Magnetic nanoparticles: recent advances in synthesis, self-assembly and applications. Journal of Materials Chemistry, 21(42) (2011) 16819-16845. https://doi.org/10.1039/c1jm11845e

[11] A.-H. Lu, W. Schmidt, N. Matoussevitch, H. Bönnemann, B. Spliethoff, B. Tesche, E. Bill, W. Kiefer, F. Schüth, Nanoengineering of a Magnetically Separable Hydrogenation Catalyst, Angewandte Chemie International Edition, 43 (33) (2004) 4303-4306. https://doi.org/10.1002/anie.200454222

[12] A. Aseri, S.K. Garg, A. Nayak, S.K. Trivedi, J. Ahsan, Magnetic Nanoparticles: Magnetic Nano-technology Using Biomedical Applications and Future Prospects, Int. J. Pharm. Sci. Rev. Res., 31(2) (2015) 119-131.

[13] A. K. Gupta, M. Gupta, Synthesis and surface engineering of iron oxide nanoparticles for biomedical applications, Biomaterials, 26(18) (2005) 3995-4021. https://doi.org/10.1016/j.biomaterials.2004.10.012

[14] B. Ramaswamy, S.D. Kulkarni, P.S. Villar, R.S. Smith, C. Eberly, R.C. Araneda, D.A. Depireux, B. Shapiro, Movement of magnetic nanoparticles in brain tissue: mechanisms and safety, Nanomedicine: Nanotechnology, Biology and Medicine, 11(7) (2015) 1821-9. https://doi.org/10.1016/j.nano.2015.06.003

[15] T. An, J. Chen, X. Nie, G. Li, H. Zhang, X. Liu, H. Zhao, Synthesis of Carbon Nanotube-Anatase TiO_2 Sub-micrometer-sized Sphere Composite Photocatalyst for Synergistic Degradation of Gaseous Styrene. ACS applied materials & interfaces, 4(11) (2012) 5988-5996. https://doi.org/10.1021/am3016476

[16] H. Teymourian, A. Salimi, S. Khezrian, $Fe3O4$ magnetic nanoparticles/reduced graphene oxide nanosheets as a novel electrochemical and bioeletrochemical sensing platform, Biosensors and Bioelectronics, 49 (2013) 1-8. https://doi.org/10.1016/j.bios.2013.04.034

[17] B. Zhang, Y. Du, P. Zhang, H. Zhao, L. Kang, X. Han, P. Xu, Microwave absorption enhancement of $Fe3O4$/polyaniline core/shell hybrid microspheres with controlled shell thickness. Journal of Applied Polymer Science, 130 (3) (2013) 1909-1916. https://doi.org/10.1002/app.39332

[18] V. Philip, V. Mahendran, L.J. Felicia, A Simple, In-Expensive and Ultrasensitive Magnetic Nanofluid Based Sensor for Detection of Cations, Ethanol and Ammonia. J. Nanofluids, 2(2) (2013) 112-119. https://doi.org/10.1166/jon.2013.1050

[19] S. Mornet, S. Vasseur, F. Grasset, P. Veverka, G. Goglio, A. Demourgues, J. Portier, E. Pollert, E. Duguet, Magnetic nanoparticle design for medical applications, Progress in Solid State Chemistry. 34 (2-4) (2006) 237-247. https://doi.org/10.1016/j.progsolidstchem.2005.11.010

[20] B. Gleich, J. Weizenecker, Tomographic imaging using the nonlinear response of magnetic particles, Nature, 435 (7046) (2005) 1214-1217. https://doi.org/10.1038/nature03808

[21] N.A. Frey, S. Peng, K. Chenga, S. Sun, Magnetic nanoparticles: synthesis, functionalization, and applications in bioimaging and magnetic energy storage, Chemical Society Reviews, 38(9) (2009) 2532-2542. https://doi.org/10.1039/b815548h

[22] Hyeon, Taeghwan Chemical synthesis of magnetic nanoparticles, Chemical Communications, (8) (2003) 927-934. https://doi.org/10.1039/b207789b

[23] J. Philip, P.D. Shima, B. Raj, Nanofluid with tunable thermal properties, Applied Physics Letters, 92 (4) (2006) 043108. https://doi.org/10.1063/1.2838304

[24] V. Chaudhary, Z. Wang, A. Ray, I. Sridhar, R.V. Ramanujan, Self pumping magnetic cooling, J. Phys D: Appl. Phys. 50(3) (2017) 03LT03. https://doi.org/10.1088/1361-6463/aa4f92

[25] J. Philip, T.J. Kumar, P. Kalyanasundaram, B. Raj, Tunable Optical Filter, Measurement Science and Technology, 14(8) (2003) 1289-1294. https://doi.org/10.1088/0957-0233/14/8/314

[26] V. Mahendran, Nanofluid based opticalsensor for rapid visual inspection of defects in ferromagnetic materials, Appl. Phys. Lett. 100(7) 073104 (2012). https://doi.org/10.1063/1.3684969

[27] V. Chaudhary, R.V. Ramanujan, Magnetocaloric Properties of Fe-Ni-Cr Nanoparticles for Active Cooling, Scientific Reports, 6 (2016) 351-356 https://doi.org/10.1038/srep35156

[28] V. Chaudhary, X. Chen, R.V. Ramanujan, Iron and manganese based magnetocaloric materials for near room temperature thermal management, Progress in Materials Science, 100 (2019) 64-98. https://doi.org/10.1016/j.pmatsci.2018.09.005

[29] D.W. Elliott, W. Zhang, Field Assessment of Nanoscale Bimetallic Particles for Groundwater Treatment". Environmental Science and Technology, 35(24) (2001): 4922-4926. https://doi.org/10.1021/es0108584

[30] T. Yoon, J. Kim, J. Kim, J.K. Lee, Electrostatic Self-Assembly of Fe3O4 Nanoparticles on Graphene Oxides for High Capacity Lithium-Ion Battery Anodes. Energies, 6(9) (2013) 4830-4840. https://doi.org/10.3390/en6094830

[31] G. Zhou, D.W. Wang, F. Li, L. Zhang, Graphene-wrapped Fe3O4 anode material with improved reversible capacity and cyclic stability for lithium ion batteries. Chemistry of materials, 22(18) (2010) 5306-5313. https://doi.org/10.1021/cm101532x

[32] H.L. Xu, Y. Shen, H. Bi, Reduced Graphene Oxide Decorated with Fe3O4 Nanoparticles as High Performance Anode for Lithium Ion Batteries. Key Engineering Materials, 519 (2012) 108-112. https://doi.org/10.4028/www.scientific.net/KEM.519.108

[33] https://tr.wikipedia-on-ipfs.org/wiki/Li_ion.html

[34] J. Lopez, M. Godlez, JC. Viera, C. Blanco, Fast-Charge İn Lithium-Ion Batteries For Portable Applications, Telecommunications Energy Conference, Sept. 19-23, (2004) 19-24.

[35] https://www.elektrikport.com/universite/samsung-cinde-lityum-iyon-pil-fabrikasi-acti/16817#ad-image-0

[36] D. Linden, T.B. Reddy, Handbook of batteries, 3rd ed. McGraw-Hill, New York, (2001).

[37] Battery University since 2003, Charging Li-ion, http://batteryuniversity.com/learn/article/charging_lithium_ion_batteries, 21.04.2019.

[38] M. Uysal, H. Gül, Characterization and Electrochemical Properties of the Sn-Cu/Rgo (Reduced Graphene Oxide) Anode Materials for Lithium Ion Batteries, APJES, 5-3 (2017) 19-25.

[39] M. Alaf, D. Gültekin, H. Akbulut, Double phase tinoxide/tin/MWCNT nanocomposite negative electrodes for lithium microbatteries, Microelectronic Engineering, 126 (2014) 143-147. https://doi.org/10.1016/j.mee.2014.06.029

[40] S. Goriparti, E. Miele, F.D. Angelis, E.D. Fabrizio, R.P. Zaccaria, C. Capiglia, Review on recent progress of nanostructured anode materials for Li-ion batteries, Journal of Power Sources, 257 (2014) 421-443. https://doi.org/10.1016/j.jpowsour.2013.11.103

[41] B. Scrosati, J. Garche, Lithium batteries: Status, prospects and future, Journal of Power Sources, 195 (2010) 2419-2430. https://doi.org/10.1016/j.jpowsour.2009.11.048

[42] O. Mao, R.L. Turnerb, I.A. Courtneya, B.D. Fredericksen, M.I. Buckett, L.J. Krause, J.R. Dahn, Active/inactive nanocomposites as anodes for Li-Ion batteries, Journal of Electrochem. Society, 2 (1999) 3-5. https://doi.org/10.1149/1.1390715

[43] L. Xian-Ming, H. Zhen Dong, W.O. Sei, Z. Biao, M. Peng-C, M.F.Y. Matthew, K. Jang-K, Carbon nanotube (CNT)-based composites as electrode material for rechargeable Li-ion batteries: A review, Composites Science and Technology, 72 (2012) 121-144. https://doi.org/10.1016/j.compscitech.2011.11.019

[44] Z. Junsheng, W. Dianlong, L. Tiefeng, G. Chenfeng, Preparation of Sn-Co-graphene composites with superior lithium storage capability, Electrochimica Acta, 125 (2014) 347-353. https://doi.org/10.1016/j.electacta.2014.01.122

[45] R. Marom, S.F. Amalraj, N. Leifer, D. Jacob, D. Aurbach, A review of advanced and practical lithium battery materials, Journal of Materials. Chemistry, 21 (2011) 9938-9954. https://doi.org/10.1039/c0jm04225k

[46] M.O. Güler, The Production and characterization of tin (II) oxide composite anode electrodes for lithium ion batteries, Sakarya Üniversitesi Fen Bilimleri Enstitüsü Dergisi, 21(2) (2017) 150-156. https://doi.org/10.16984/saufenbilder.296995

[47] R. Zhang, J.Y. Lee, Z.L. Liu, Pechini process-derived tin oxide and tin oxideegraphite composites for lithium-ion batteries, J. Power Sources, 112 (2012) 596-605. https://doi.org/10.1016/S0378-7753(02)00483-4

[48] G. Derrien, J. Hassoun, S. Panero, B. Scrosati, Nanostructured Sn-C composite as an advanced anode material in high-performance lithium-ion batteries, Advanced Materials 19(17) (2007) 2336-2340. https://doi.org/10.1002/adma.200700748

[49] S. Sharma, L. Fransson, E. Sjostedt, L. Nordstrom, B. Johansson, K. Edstrom, A theoretical and experimental study of the lithiation of N-Cu6Sn5 in a lithium-ion battery, J Electrochem Soc., 150 (2003) 330-334. https://doi.org/10.1149/1.1544634

[50] X.W. Lou, C.M. Li, L.A. Archer, Designed synthesis of coaxial SnO2@carbon hollow nanospheres for highly reversible lithium storage, Advanced Materials, 21 (2009) 2536-2539. https://doi.org/10.1002/adma.200803439

[51] X. Li, X. Meng, J. Liu, D. Geng, Y. Zhang, M. Banis, Y. Li, R. Li, X. Sun, M. Cai, M. Verbrugge, Tin oxide with controlled morphology and crystallinity by atomic layer deposition onto graphene nanosheets for enhanced lithium storage, Advanced Functional Materials, 22(8) (2012) 1647-1654. https://doi.org/10.1002/adfm.201101068

[52] M. Valvo, U. Lafont, L. Simonin, E.M. Kelder, Sn-Co compound for Li-ion battery made via advanced electrospraying, Journal of Power Sources, 174 (2007) 428-434. https://doi.org/10.1016/j.jpowsour.2007.06.156

[53] Z. Chen, J. Qian, X. Ai, Y. Cao, H. Yan, Preparation and electrochemical performance of Sn-Co-C composite as anode material for Li-ion batteries, Journal of Power Sources, 189 (2009) 730-732. https://doi.org/10.1016/j.jpowsour.2008.08.027

[54] X. Li, Y. Zhong, M. Cai, M.P. Balogh, D. Wang, Y. Zhang, R. Li, X. Sun, Tin-alloy heterostructures encapsulated in amorphous carbon nanotubes as hybrid anodes in rechargeable lithium ion batteries, Electrochimica Acta, 89 (2013) 387-393. https://doi.org/10.1016/j.electacta.2012.11.097

[55] X.M. Liu, Z. Huang, S. Oh, B. Zhang, P.C. Ma, M. M.F. Yuen, J.K. Kim, Carbon nanotube (CNT)-based composites as electrode material for rechargeable Li-ion batteries: A review, Composites Science and Technology, 72 (2012) 121-144. https://doi.org/10.1016/j.compscitech.2011.11.019

[56] B. Xu, D. Qian, Z. Wang, Y.S. Meng, Recent progress in cathode materials research for advanced lithium ion batteries Materials Science and Engineering: R: Reports, 73(5-6) (2012) 51-65. https://doi.org/10.1016/j.mser.2012.05.003

[57] A.H. Whitehead, J.M. Elliott, J.R. Owen, Nanostructured tin for use as a negative electrode material in Li-ion. J. Power Sources, 81-82 (1999) 33-38. https://doi.org/10.1016/S0378-7753(99)00126-3

[58] Y.H. Jin, K.M. Min, H.W. Shim, S.D. Seo, I.S. Hwang, K.S. Park, D.W. Kim, Facile synthesis of nano-Li4 Ti5O12 for high-rate Li-ion battery anodes, Nanoscale Research Letters, 7(10) (2012) 1-6. https://doi.org/10.1186/1556-276X-7-10

[59] C. Dewan, D. Teeters, Vanadia xerogel nanocathodes used in lithium microbatteries. J. Power Sources, 119-121 (2003) 310-315. https://doi.org/10.1016/S0378-7753(03)00165-4

[60] H. Yan, S. Sokolov, J.C. Lytle, A. Stein, F. Zhang, W.H. Smyrl, Colloidal-crystal-templated synthesis of ordered macroporous electrode materials for Lithium secondary batteries, J. Electrochem. Soc. 150(8) (2003) 1102-1107. https://doi.org/10.1149/1.1590324

[61] P.L. Taberna, S. Mitra, P. Poizot, P. Simon, J.-M. Tarascon, High rate capabilities Fe3O4-based Cu nano-architectured electrodes for lithium-ion battery applications, Nature Materials, 5 (2006) 567-573. https://doi.org/10.1038/nmat1672

[62] A. Manthiram, A.V. Murugan, A. Sarkar, T. Muraliganth, Nanostructured electrode materials for electrochemical energy storage and conversion, Energy Environ. Sci., 1(6) (2008) 621-638. https://doi.org/10.1039/b811802g

[63] R.A. Huggins, W.D. Nix, Decrepitation model for capacity loss during cycling of alloys in rechargeable electrochemical systems, Ionics, 6 (2000) 57-63. https://doi.org/10.1007/BF02375547

[64] Y. Zhang, Y. Liu, M. Liu, Nanostructured Columnar Tin Oxide Thin Film Electrode for Lithium Ion Batteries, Chem. Mater., 18 (2006) 4643-4646. https://doi.org/10.1021/cm0519378

[65] Y.H. Chen, C.W. Wang, G. Liu, X.Y. Song, V.S. Battaglia, A.M. Sastry, Selection of Conductive Additives in Li-Ion Battery Cathodes, J. Electrochem. Soc., 154(10) (2007) 978-986. https://doi.org/10.1149/1.2767839

[66] R. Teki, M.K. Datta, R. Krishnan, T.C. Parker, T.M. Lu, P.N. Kumta, N. Koratkar, Nanostructured silicon anodes for lithium ion rechargeable batteries, Small, 5(20) (2009) 2236-2242. https://doi.org/10.1002/smll.200900382

[67] Y. Zhao, G. Liu, L. Liu, Z.J. Jiang, High-performance thin-film Li4Ti5O12 electrodes fabricated by using ink-jet printing technique and their electrochemical properties, Solid State Electrochem., 13 (2009) 705-711. https://doi.org/10.1007/s10008-008-0575-6

[68] G. Venugopal, A. Hunt, F. Alamgir, Nanomaterials for Energy Storage in Lithium-ion Battery Applications, Material Matters, (2015) 42-45.

Materials Research Forum LLC
https://doi.org/10.21741/9781644900918-5

[69] J.B. Goodenough, Energy storage materials: A perspective, Energy Storage Materials, 1 (2015) 158-161. https://doi.org/10.1016/j.ensm.2015.07.001

[70] J. Lang, L. Qi, Y. Luo, H. Wu, High performance lithium metal anode: Progress and prospects, Energy Storage Materials, 7 (2017) 115-129. https://doi.org/10.1016/j.ensm.2017.01.006

[71] J.B. Goodenough, K.S. Park, The Li-ion rechargeable battery: a perspective, Journal of the American Chemical Society, 135(4) (2013) 1167-1176. https://doi.org/10.1021/ja3091438

[72] J.B. Goodenough, Y. Kim, Challenges for Rechargeable Li Batteries, Chemistry of Materials, 22(3) (2010) 587-603. https://doi.org/10.1021/cm901452z

[73] G.G. Wallace, J. Chen, A.J. Mozer, M. Forsyth, D.R. MacFarlane, C. Wang, Nanoelectrodes: energy conversion and storage. Mater Today, 12 (2009) 20-27. https://doi.org/10.1016/S1369-7021(09)70177-4

[74] H. Li, Z. Wang, L. Chen, X. Huang, Research on advanced materials for Li-ion batteries, Adv. Mater. 21 (2009) 4593-4607. https://doi.org/10.1002/adma.200901710

[75] M.S. Whittingham, Lithium batteries and cathode materials. Chem. Rev. 104 (2004) 4271-4301. https://doi.org/10.1021/cr020731c

[76] C. Jiang, E. Hosono, H. Zhou, Nanomaterials for lithium ion batteries, Nano Today 1 (2006) 28-33. https://doi.org/10.1016/S1748-0132(06)70114-1

[77] M.S. Whittingham, Materials challenges facing electrical energy storage. MRS Bull, 33 (2008) 411-419. https://doi.org/10.1557/mrs2008.82

[78] E. Frackowiak, S. Gautier, H. Gaucher, S. Bonnamy, F. Beguin, Electrochemical storage of lithium multiwalled carbon nanotubes. Carbon, 37 (1999) 61-69. https://doi.org/10.1016/S0008-6223(98)00187-0

[79] J. Chen, A.I. Minett, Y. Liu, C. Lynam, P. Sherrell, C. Wang, Direct growth of flexible carbon nanotube electrodes. Adv Mater, 20(3) (2008) 566-570. https://doi.org/10.1002/adma.200701146

[80] A. Claye, J. Fischer, C. Huffman, A. Rinzler, R.E. Smalley, Solid-state electrochemistry of the Li single wall carbon nanotube system. J Electrochem Soc 147 (2000) 2845-2852. https://doi.org/10.1149/1.1393615

[81] P. Liu, G.L. Hornyak, A.C. Dillon, T. Gennett, M.J. Heben, J.A. Turner, Electrochemical performance of carbon nanotube materials in lithium ion batteries. J Electrochem Soc: Proc Int Symp (1999) 31-39.

[82] H.M. Hsoeh, N.H. Tai, C.Y. Lee, J.M. Chen, F.T. Wang, Electrochemical properties of the multi-walled carbon nanotube electrode for secondary lithium-ion battery. Rev Adv Mater Sci, 5 (2003) 67-71.

[83] J. Ren, Z. Wang, F. Yang, R.P. Ren, Y.K. Lv, Freestanding 3D Single-wall Carbon Nanotubes/WS2 Nanosheets Foams as Ultra-Long-Life Anodes for Rechargeable Lithium Ion Batteries, Electrochimica Acta, 267 (2018) 133-140. https://doi.org/10.1016/j.electacta.2018.01.167

[84] F. Kayış, Production of Sn-Co / CNT Composite Anodes by Pulse Current Method for Lithium Ion Batteries, Bilecik Şeyh Edebali Üniversitesi Fen Bilimleri Dergisi, 3(2) (2016) 25-29.

[85] X. Huang, H. Li, Nanometer Anode Materials for Li-Ion Batteries, Nanomaterials for Lithium-Ion Batteries: Fundamentals and Applications, (2014) 167-197. https://doi.org/10.1201/b15488-6

[86] M.D. Bhatt, J.Y. Lee, High capacity conversion anodes in Li-ion batteries: A review, International Journal of Hydrogen Energy, 44(21) (2019) 10852-10905. https://doi.org/10.1016/j.ijhydene.2019.02.015

[87] P. Li, G. Zhao, X. Zheng, X. Xu, C. Yao, W. Sun, S.X. Dou, Recent progress on silicon-based anode materials for practical lithium-ion battery applications, Energy Storage Materials, 15 (2018) 422-446. https://doi.org/10.1016/j.ensm.2018.07.014

[88] W. Li, X. Li, J. Yu, J. Liao, B. Zhao, L. Huang, A. Abdelhafiz, H. Zhang, J.H. Wang, Z. Guo, M. Liu, A self-healing layered GeP anode for high-performance Li-ion batteries enabled by low formation energy, Nano Energy, (2019) In Press, Accepted Manuscript, DOI: 10.1016/j.nanoen.2019.04.080.

[89] H. Liu, Q. Liu, Z. Yang, Porous micro/nano Li2CeO3 with baseball morphology as anode material for high power lithium ions batteries, Solid State Ionics, 334 (2019) 82-86. https://doi.org/10.1016/j.ssi.2019.02.008

[90] J. Yan, J. Yao, Z. Zhang, Y. Li, S. Xiao, 3D hierarchical porous ZnFe2O4 nano/micro structure as a high-performance anode material for lithium-ion batteries, Materials Letters, 245 (2019) 122-125. https://doi.org/10.1016/j.matlet.2019.02.113

[91] A. Sun, H. Zhong, X. Zhou, J. Tang, M. Jia, F. Cheng, Q. Wang, J. Yang, Scalable synthesis of carbon-encapsulated nano-Si on graphite anode material with high cyclic stability for lithium-ion batteries, Applied Surface Science, 470 (2019) 454-461. https://doi.org/10.1016/j.apsusc.2018.11.117

[92] S. Bao, J. Li, Y. Xiao, P. Li, L. Liu, B. Yue, Y. Li, W. Sun, W. Zhang, L. Zhang, X. Lai, In-situ porous nano-Fe3O4/C composites derived from citrate precursor as anode materials for lithium-ion batteries, Materials Chemistry and Physics, 225 (2019) 379-383. https://doi.org/10.1016/j.matchemphys.2018.12.072

[93] B. Cheng, L. Luo, H. Zhuo, S. Chen, X. Zeng, Preparation of nano-VBO3 on graphene as anode material for lithium-ion batteries, Materials Letters, 241 (2019) 60-63. https://doi.org/10.1016/j.matlet.2019.01.014

[94] S. Jia, Y. Wang, X. Liu, S. Zhao, W. Zhao, Y. Huang, Z. Li, Z. Lin, Hierarchically porous CuO nano-labyrinths as binder-free anodes for long-life and high-rate lithium ion batteries, Nano Energy, 59 (2019) 229-236. https://doi.org/10.1016/j.nanoen.2019.01.081

[95] Y. Wang, Y. Zhang, Y. Peng, H. Li, J. Li, B.J. Hwang, J. Zhao, Physical confinement and chemical adsorption of porous C/CNT micro/nano-spheres for CoS and Co9S8 as advanced lithium batteries anodes, Electrochimica Acta, 299 (2019) 489-499. https://doi.org/10.1016/j.electacta.2018.11.138

[96] G.C. Shivaraju, C. Sudakar, A.S. Prakash, High-rate and long-cycle life performance of nano-porous nano-silicon derived from mesoporous MCM-41 as an anode for lithium-ion battery, Electrochimica Acta, 294 (2019) 357-364. https://doi.org/10.1016/j.electacta.2018.10.122

[97] S. Yang, Y. Huang, D. Zhang, G. Han, Y. Cao, J. Liu, Fabrication and characterization of dinickel orthosilicate nanosheets as high performance anode material for lithium-ion batteries, Journal of Alloys and Compounds, 785 (2019) 80-88. https://doi.org/10.1016/j.jallcom.2019.01.195

[98] M. Krajewski, P.H. Lee, S.H Wu, K. Brzozka, A. Malolepszy, L. Stobinski, M. Tokarczyk, G. Kowalski, D. Wasik, Nanocomposite composed of multiwall carbon nanotubes covered by hematite nanoparticles as anode material for Li-ion batteries, Electrochimica Acta, 228 (2017) 82-90. https://doi.org/10.1016/j.electacta.2017.01.051

[99] J. Xu, Z. Han, J. Wu, K. Song, J. Wu, H. Gao, Y. Mi, Synthesis and electrochemical performance of vertical carbon nanotubes on few-layer graphene as an anode material for Li-ion batteries, Materials Chemistry and Physics, 205 (2018) 359-365. https://doi.org/10.1016/j.matchemphys.2017.11.039

[100] J. Li, Y. Li, Q. Lan, Z. Yang, X.J. Lv, Multiple phase N-doped TiO2 nanotubes/TiN/graphene nanocomposites for high rate lithium ion batteries at low

Materials Research Forum LLC
https://doi.org/10.21741/9781644900918-5

temperature, Journal of Power Sources, 423 (2019) 166-173.
https://doi.org/10.1016/j.jpowsour.2019.03.060

[101] Y.K. Sun, Cycling behaviour of LiCoO2 cathode materials prepared by PAA-assisted Sol-Gel method for rechargeable lithium batteries, J. Power Sources, 83(1-2) (1999) 223-226. https://doi.org/10.1016/S0378-7753(99)00280-3

[102] M.Y. Song, R. Lee, Synthesis by Sol-Gel method and electrochemical properties of LiNiO2 cathode material for lithium secondary battery, J. Power Sources, 111(1) (2002) 93-103. https://doi.org/10.1016/S0378-7753(02)00263-X

[103] T. Tanaka, K. Ohta, N. Arai, Year 2000 R&D status of large-scale lithium ion secondary batteries in the national project of Japan, J. Power Sources, 97-98 (2001) 2-6. https://doi.org/10.1016/S0378-7753(01)00502-X

[104] S. Tao, Q. Wu, Z. Zhan, G. Meng, Preparation of LiMO2 (M=Co, Ni) cathode materials for intermediate temperature fuel cells by sol-gel processes, Solid State Ionics, 124(1-2) (1999) 53-59. https://doi.org/10.1016/S0167-2738(99)00137-X

[105] H.J. Kweon, G.B. Kim, H.S. Lim, S.S. Nam, D.G. Park, Synthesis of LixNi0.85Co0.15O2 by the PVA-Precursor method and charge-discharge characteristics of a lithium ion battery using this material as cathode, J. Power Sources, 83(1-2) (1999) 84-92. https://doi.org/10.1016/S0378-7753(99)00271-2

[106] S.C. Park, Y.M. Kim, Y.M. Kang, K.T. Kim, P.S. Lee, J.Y. Lee, Improvement of the Rate Capability of LiMn2O4 by Surface Coating with LiCoO2, J. Power Sources, 103(1) (2001) 86-92. https://doi.org/10.1016/S0378-7753(01)00832-1

[107] T. Takamura, Trends in Advanced Batteries and Key Materials in the New Century, Solid State Ionics, 152-153 (2002) 19-34. https://doi.org/10.1016/S0167-2738(02)00325-9

[108] E. Özçelik, G. Özkan, Synthesis and characterization of LiCoO2 used as cathode material in secondary lithium batteries, J. Fac. Eng. Arch. Gazi Univ. 21(3) (2006) 423-425.

[109] D. Ni, W. Sun, L. Xie, Q. Fan, Z. Wang, K. Sun, Bismuth oxyfluoride @ CMK-3 nanocomposite as cathode for lithium ion batteries, Journal of Power Sources, 374 (2018) 166-174. https://doi.org/10.1016/j.jpowsour.2017.11.017

[110] T. Cetinkaya, H. Akbulut, M. Tokur, S. Ozcan, M. Uysal, High capacity Graphene/α-MnO2 nanocomposite cathodes for Li-O2 batteries, International Journal of Hydrogen Energy, 41(23) (2016) 9746-9754. https://doi.org/10.1016/j.ijhydene.2016.02.093

[111] C. Luo, H. Sun, Z. Jiang, H. Guo, M. Gao, M. Wei, Z. Jiang, H. Zhou, S.G. Sun, Electrocatalysts of Mn and Ru oxides loaded on MWCNTS with 3D structure and synergistic effect for rechargeable Li-O2 battery, Electrochimica Acta, 282 (2018) 56-63. https://doi.org/10.1016/j.electacta.2018.06.040

[112] M. Haris, S. Atiq, S.M. Abbas, A. Mahmood, S.M. Ramay, S. Naseem, Acetylene black coated V2O5 nanocomposite with stable cyclability for lithium-ion batteries cathode, Journal of Alloys and Compounds, 732 (2018) 518-523. https://doi.org/10.1016/j.jallcom.2017.10.221

[113] H. Yin, X.X. Yu, Y.W. Yu, M L. Cao, H. Zhao, C. Li, M.Q. Zhu, Tellurium nanotubes grown on carbon fiber cloth as cathode for flexible all-solid-state lithium-tellurium batteries, Electrochimica Acta, 282 (2018) 870-876. https://doi.org/10.1016/j.electacta.2018.05.190

[114] D. Tang, W. Zhang, Z.A Qiao, Y. Liu, D. Wang, Polyanthraquinone/CNT nanocomposites as cathodes for rechargeable lithium ion batteries, Materials Letters, 214 (2018) 107-110. https://doi.org/10.1016/j.matlet.2017.11.119

[115] N. Kumar, J. R. Rodriguez, V. G. Pol, A. Sen, Facile synthesis of 2D graphene oxide sheet enveloping ultrafine 1D LiMn2O4 as interconnected framework to enhance cathodic property for Li-ion battery, Applied Surface Science, 463 (2019) 132-140. https://doi.org/10.1016/j.apsusc.2018.08.210

[116] S. Liang, M. Qin, J. Liu, Q. Zhang, T. Chen, Y. Tang, W. Wang, Facile synthesis of multiwalled carbon nanotube-LiV3O8 nanocomposites as cathode materials for Li-ion batteries, Materials Letters, 93 (2013) 435-438. https://doi.org/10.1016/j.matlet.2012.09.071

[117] B. Sen, S. Kuzu, E. Demir, S. Akocak F. Sen, Highly monodisperse RuCo nanoparticles decorated on functionalized multiwalled carbon nanotube with the highest observed catalytic activity in the dehydrogenation of dimethylamine−borane, International Journal of Hydrogen Energy, 42 (2017) 23292-23298. https://doi.org/10.1016/j.ijhydene.2017.06.032

[118] R. Ayrancı, B. Demirkan, B. Sen, A. Savk, M. Ak, F. Sen, Use of the monodisperse Pt/Ni@rGO nanocomposite synthesized by ultrasonic hydroxide assisted reduction method in electrochemical nonenzymatic glucose detection, Materials Science and Engineering: C, 99 (2019) 951-956. https://doi.org/10.1016/j.msec.2019.02.040

[119] B. Sen, A. Savk, F. Sen, Highly Efficient Monodisperse Pt Nanoparticles Confined in The Carbon Black Hybrid Material for Hydrogen Liberation, Journal of Colloid and Interface Science, 520 (2018) 112-118. https://doi.org/10.1016/j.jcis.2018.03.004

[120] S. Ertan, F. Sen, S. Sen G. Gokagac, Platinum nanocatalysts prepared with different surfactants for C1-C3 alcohol oxidations and their surface morphologies by AFM, Journal of Nanoparticle Research, 14 (2012) 922-934. https://doi.org/10.1007/s11051-012-0922-5

[121] B. Sen, B. Demirkan, A. Savk, R. Kartop, M.S. Nas, M.H. Alma, S. Sürdem, F. Sen, High-performance graphite-supported ruthenium nanocatalyst for hydrogen evolution reaction, Journal of Molecular Liquids, 268 (2018) 807-812. https://doi.org/10.1016/j.molliq.2018.07.117

[122] R. Ayranci, G. Baskaya, M. Guzel, S. Bozkurt, M. Ak, A. Savk, F. Sen, Activated Carbon Furnished Monodisperse Pt Nanocomposites as a Superior Adsorbent for Methylene Blue Removal from Aqueous Solutions, Nano-Structures and Nano-Objects, 17 (2017) 4799-4804. https://doi.org/10.1166/jnn.2017.13776

[123] R. Ayranci, G. Baskaya, M. Guzel, S. Bozkurt, M. Ak, A. Savk, F. Sen, Enhanced optical and electrical properties of PEDOT via nanostructured carbon materials: A comparative investigation, Nano-Structures & Nano-Objects, 11 (2017) 13-19. https://doi.org/10.1016/j.nanoso.2017.05.008

[124] B. Sen, A. Aynur, T.O. Okyay, A. Savk, R. Kartopu, F. Sen, Monodisperse palladium nanoparticles assembled on graphene oxide with the high catalytic activity and reusability in the dehydrogenation of dimethylamine-borane, International Journal of Hydrogen Energy, 43 (2018) 20176-20182. https://doi.org/10.1016/j.ijhydene.2018.03.175

[125] F. Sen, Y. Karatas, M. Gulcan, M. Zahmarikan, Amylamine stabilized platinum(0) nanoparticles: active and reusable nanocatalyst in the room temperature dehydrogenation of dimethylamine-borane, RSC Adv. 4 (2014) 1526-1531. https://doi.org/10.1039/C3RA43701A

[126] Y. Koskun, A. Savk, B. Sen, F. Sen, Highly Sensitive Glucose Sensor Based on Monodisperse Palladium Nickel/Activated Carbon Nanocomposites, Analytica Chimica Acta, 1010 (2018) 37-43. https://doi.org/10.1016/j.aca.2018.01.035

[127] B. Sahin, E. Demir, A. Aynur, H. Gunduz, F. Sen, Investigation of The Effect Of Pomegranate Extract And Monodisperse Silver Nanoparticle Combination on MCF-7 Cell Line, Journal of Biotechnology, 260 (2017) 79-83. https://doi.org/10.1016/j.jbiotec.2017.09.012

[128] B. Sen, E. Kuyuldar, B. Demirkan, T.O. Okyay, A. Savk, F. Sen, Highly Efficient Polymer Supported Monodisperse Ruthenium-nickel Nanocomposites for

Dehydrocoupling of Dimethylamine Borane, Journal of Colloid and Interface Science, 526 (2018) 480-486. https://doi.org/10.1016/j.jcis.2018.05.021

[129] N. Lolak, E. Kuyuldar, H. Burhan, H. Goksu, S. Akocak, F. Sen, Composites of Palladium-Nickel Alloy Nanoparticles and Graphene Oxide for the Knoevenagel Condensation of Aldehydes with Malononitrile, ACS Omega, 4 (2019) 6848-6853. https://doi.org/10.1021/acsomega.9b00485

[130] H. Goksu, B. Çelik, Y. Yunus, F. Sen, B. Kilbas, Superior Monodisperse CNT-Supported CoPd (CoPd@CNT) Nanoparticles for Selective Reduction of Nitro Compounds to Primary Amines with NaBH 4 in Aqueous Medium, ChemistrySelect, 1 (2016) 2366-2372. https://doi.org/10.1002/slct.201600509

[131] F. Sen, H. Gokagac, Different Sized Platinum Nanoparticles Supported on Carbon: An XPS Study on These Methanol Oxidation Catalysts, Journal of Physical Chemistry C, 111 (2007) 5715-5720. https://doi.org/10.1021/jp068381b

[132] S. Gunbatar, A. Aygun, Y. Karatas, M. Gulcan, F. Sen, Carbon-nanotube-based Rhodium Nanoparticles as Highly-Active Catalyst for Hydrolytic Dehydrogenation of Dimethylamineborane at Room Temperature, Journal of Colloid and Interface Science, 530 (2018) 321-327. https://doi.org/10.1016/j.jcis.2018.06.100

[133] E. Demir, A. Savk, B. Sen, F. Sen, A Novel Monodisperse Metal Nanoparticles Anchored Graphene Oxide as Counter Electrode for Dye-Sensitized Solar Cells, Nano-Structures and Nano-Objects, 12 (2017) 41-45. https://doi.org/10.1016/j.nanoso.2017.08.018

[134] B. Sen, A. Aygun, A. Savk, S. Akocak, F. Sen, Bimetallic Palladium-iridium Alloy Nanoparticles as Highly Efficient and Stable Catalyst for The Hydrogen Evolution Reaction, International Journal of Hydrogen Energy, 43 (2018) 20183-20191. https://doi.org/10.1016/j.ijhydene.2018.07.081

[135] Y. Yıldız, S. Kuzu, B. Sen, A. Savk, S. Akocak, F. Sen, Different Ligand Based Monodispersed Pt Nanoparticles Decorated with rGO As Highly Active and Reusable Catalysts for The Methanol Oxidation, International Journal of Hydrogen Energy, 42 (2017) 13061-13069. https://doi.org/10.1016/j.ijhydene.2017.03.230

Lithium-ion Batteries - Materials and Applications
Materials Research Foundations **80** (2020) 148-160

Materials Research Forum LLC
https://doi.org/10.21741/9781644900918-6

Chapter 6

Recent Advances in Nanomaterials for Li-ion Batteries

K. Chandra Babu Naidu[1*], N. Suresh Kumar[2], Rajender Boddula[3], S. Ramesh[1],
Ramyakrishna Pothu[4], Prasun Banerjee[1], M.S.S.R.K.N. Sarma[1], H. Manjunatha[5], B. Kishore[6]

[1]Dept. of Physics, GITAM Deemed to be University, Bangalore -562163, India

[2]Dept. of Physics, JNTUA, Anantapuramu-515002, A.P, India

[3]CAS Key Laboratory of Nanosystem and Hierarchical Fabrication, National Center for Nanoscience and Technology, Beijing 100190, PR China

[4]College of Chemistry and Chemical Engineering, Hunan University, Changsha 410082, PR China.

[5]Dept.of Chemistry, GITAM Deemed to be University, Bangalore -562163, India

[6]Dept. of Mechanical Engineering, GITAM Deemed to be University, Bangalore - 562163, India

*chandrababu954@gmail.com

Abstract

This chapter provides the information about different materials for the fabrication of Li-ion batteries. Further, the basic structure of Li-ion batteries is discussed. Subsequently, the electrochemical energy storage efficiency of various Li-ion batteries is described as a function of distinct electrochemical parameters.

Keywords

Li-ion batteries, Electrochemical properties, Advanced materials, Specific capacitance

Contents

Materials Research Forum LLC
https://doi.org/10.21741/9781644900918-6

1. Introduction

Rapidly growing technology of the current world requires the electrical energy storage for systematic run of digital electronic devices (DED), hybrid electrical vehicles (HEV), and aerospace vehicles. For this purpose, the most commonly used electrical energy storage devices are capacitor, battery, pumped hydropower, supercapacitor, flywheels, and compressed air [1]. Out of these energy storage elements, the batteries have good recognition owing to their efficient power density, thermal stability and energy density parameters. This efficiency of various electrochemical parameters mainly depends on the battery chemistry as well as its design [1]. However, in case of Li-ion batteries, the life-extension, inexpensive, energy density, cyclic retention and charging speed are more advantageous than the other batteries/electrical energy storage devices [2]. Therefore, these Li-ion batteries are quite useful in the computers, smart mobiles, power tools, grid energy storages, electronic machines, fitness bands, and portable electronic objects [1]. Although, the abundance of lithium is limited on the earth surface, the Li-ion batteries have acquired considerable attention in the electric vehicles, and aerospace machines [1]. It is noticed that the Li-ion batteries will have a state of charge (SOC) ranging from 20 – 80 % in order to avoid deep discharge and overcharge [2].

Munson [4] showed the voltage versus capacity trend of Li-ion batteries at different temperatures of 25 °C, 0 °C and -20 °C at a constant current of 2.3 A. This is depicted in Figure 1. It is understood that the Li-ion batteries revealed low capacity (%) at high voltages. On the other hand, the capacity is gradually increased to high percentage at the intermediate voltages (for different temperatures). Subsequently, the capacity is further enhanced to higher value of 80 – 100 % between the voltages of 2.0-2.5 V. From this, it is confirmed that the efficiency of Li-ion batteries is very much preferable at room temperature (25 °C), whereas the electrochemical performance is come down while moving towards low temperatures. However, in the current world, the usage of electric vehicles, laptops, smart mobiles, and other portable electronic objects is extensively made only at room temperature rather than the lower temperatures. Therefore, the Li-ion batteries are well suited for electrical energy storage in the above mentioned electronic devices/machines.

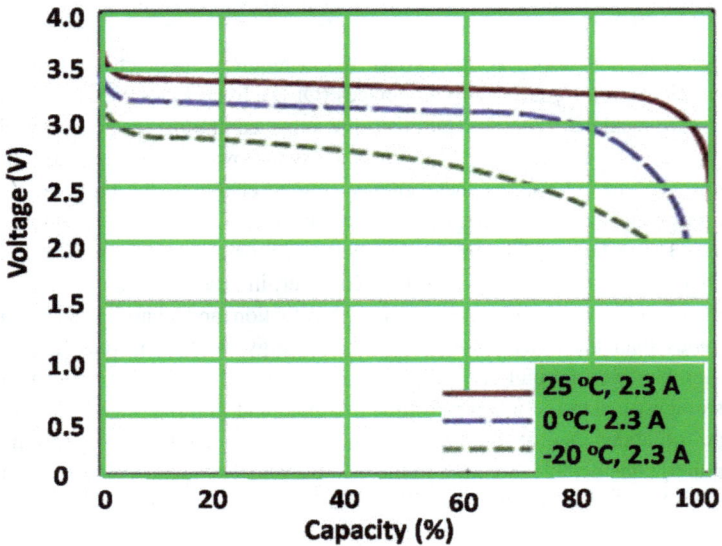

Figure 1:Li-ion battery performance (voltage versus capacity) at different temperatures.

2.   Structure and working of Li-ion battery

It is a known fact that the efficiency of electrical energy storage of batteries typically depends on the design of anode, cathode, and electrolytes. The schematic representation of the structure of Li-ion battery is provided in Figure 2 [1]. In general, the storage of electrical energy is a dependent parameter of chemical components of different electrochemical potentials within the batteries. Zhaoxiang et al. [1], considered a familiar and commercial Li-ion battery material. This material is divided into pairs called $LiCoO_2$ and graphite. In order to elucidate the charging and discharging process of a static battery, the lithium cobalt oxide and graphite materials are treated as cathode and anodes respectively. It can be obviously seen in Figure 2. In Figure 2, cathode, anode, electrolyte, separator, and current collector components are noticed. The current collector is employed in order to transport the electrons from electrodes to the terminals of battery. Likewise, the separator which reveals normally porous structure can prevent the shorting of cathode (+ve electrode) and anode (-ve electrode). In the same fashion, the electrolyte provides the transportation of ions in order to keep the constant charge during the charging or discharging process. Moreover, the carbon additive along with binder is mixed to the present electrodes of $LiCoO_2$ and graphite for making the electron transfer in a convenient way and to achieve mechanical integrity during the cell fabrication. The cathode and anode materials can be prepared using the following chemical reaction. Herein, the forward arrow mark indicates the charging while the backward arrow mark shows the discharge mechanism.

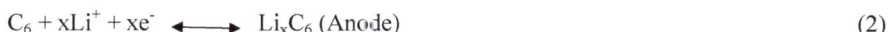

$$LiCoO_2 \longleftrightarrow Li_{1-x}CoO_2 + xLi^+ + xe^- \text{ (Cathode)} \qquad (1)$$

$$C_6 + xLi^+ + xe^- \longleftrightarrow Li_xC_6 \text{ (Anode)} \qquad (2)$$

It is clear that the lithium ions are transferred from cathode to anode during the charging process while the same ions will be transferred in opposite direction during the discharge process. The bulk form of the prepared Li-ion batteries for electric vehicles and smartphones is provided in the Figure 2.

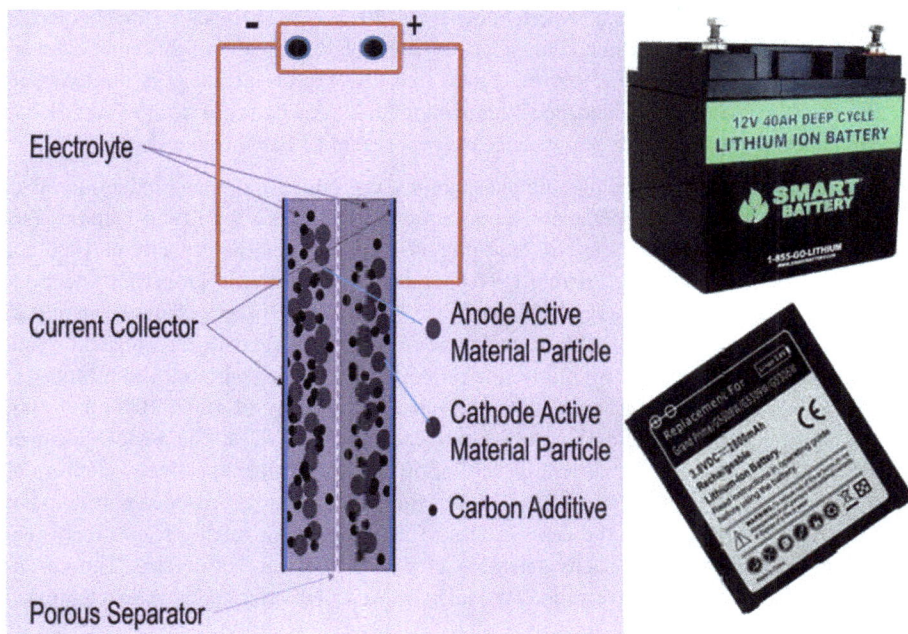

Figure 2:Structure of Li-ion battery and internal components.

3.   Electrochemical behavior of various materials for Li-ion batteries

It has been an established fact that the Li-based materials, metals, phosphates, oxides, sulfides, fluorides, thin films, silicates, metal oxides, ferrites, perovskites, nanosheets, nanorods, nanofibers, composites, polymers, nanobelts, nanoparticles, nanowires, nanoboxes, microboxes, nanotubes, etc. The electrochemical performance of each kind of material is listed in Table 1. The materials along with the electrochemical parameters are extracted from the references provided in Table 1 [5 – 36]. Among the lithium-based materials, it is noticed that the compounds such as $Li_{22}Si_5$, Li-metal, Li-ion polysulfide, Li_9Al_4,and Li_2S recorded the highest specific capacitance values of 4200, 3860, 2430, 2235 and 1672 mAh/g respectively [see Table 1]. The $Li_{22}Si_5$ showed the extreme capacity (4200 mAh/g) at very low voltage (Li/Li$^+$) of 0.4 V. However, it has one disadvantage of quick degradation of that material owing to the high expansion, and solid electrolyte interface layer growth. Likewise, Li-metal revealed high specific capacity (3860 mAh/g) at low voltage of 0.03 V. It provided the flat potential curve. Besides, the defect observed this metal has limited Coulombic efficiency. Li-ion polysulfide expressed the capacity of 2430 mAh/g. This material showed the high energy density of 450 Wh/kg with good capacity retention about 70 % for 240 cycles (for charge and discharge). Li_9Al_4 also exhibited high capacity. The big advantages of this compound are good availability, cheap and environmental free. But the disadvantage is limited cycle life. Li_2S is having high abundance by providing high capacity at 0.4 V. However, due to limited cycle life, it has less significance over the rest of Li-based compounds.

Furthermore, it is observed that the phosphorous, carbon, silicon, metal oxides and their composites showed considerable electrochemical performance for Li-ion batteries (see Table 1). In view of this, the NiP_2 and pristine C_3N acquired the capacity of 1000 and 1071 mAh/g respectively. Pristine C_3N provides good electronic properties and cycle performance due to low energy band gap of 0.39 eV and small energy barrier gap of 0.27 eV. The $SiN_{0.79}$ expressed high capacitance of 1500 mAh/g with an excellent cyclic retention for 2000 cycles. Similarly, the different forms of metal oxides like MoO_2, α-MoO_3 nanobelt (nb), and MoO_3 nb/C attributed the capacity of 1607, 1000, and 1000 mAh/g respectively. Herein, the MoO_2 has retention for 50 cycles. The α-MoO_3 nanobelt (nb) and MoO_3 nb/C offer current density of 50 and 0.1 mA/g respectively. Concerning with the composites, the MnO/rGO (graphene oxide) nanocomposite and MnO/C@MRGO attained the capacitances of 2358 and 1178 mAh/g respectively. The MnO/rGO showed the capacity retention of 570 mAh/g for 100 cycles. The current density about 100 mA/g is recorded for MnO/C@MRGO to offer good cyclic retention.

The iron oxide and its based composites revealed significant capacitances for Li-ion batteries. For instance, the Fe_2O_3/SWCNT (single-wall carbon nanotube), Fe_3O_4/N-doped graphene and 3D-α-Fe$_2$O$_3$ exhibited the specific capacitance of 1243 (at current density of 50 mA/g), 1188 (at current density of 3000 mA/g) and 1157 mAh/g respectively. Herein, the Fe$_3$O$_4$/N-doped graphene interestingly performed good capacitive retention for 1000 cycles. 3D-α-Fe$_2$O$_3$ offered cyclic retention at current density of 650μA/cm^2 for 500 cycles (0.13 % capacity fading/cycle). The iron oxide nanorods/Ti attributed 562 mAh/g with high cyclic retention for 50 cycles while the same iron oxide after doping with carbon nanotubes showed low capacitance of 375 mAh/g at current density of 3000 mA/g by possessing considerable cyclic retention for 1000 cycles. As a whole, it is confirmed that the doping of various metals and CNTs to iron oxides reduced the specific capacitance and offered good cyclic retention for larger number of cycles. In the case film-based materials like CoN film offered capacitance of 990 mAh/g when compared with other films mentioned in Table 1.

The cobalt oxides and their composites played a vital role to perform the Li-ion battery applications. Different forms of Co$_3$O$_4$ are presented in Table 1 pertaining their specific capacitances. It is found that the Fe$_2$O$_3$/Co$_3$O$_4$ double-helled hollow microcubes and nanoporous TiO$_2$/Co$_3$O$_4$composites showed the highest and lowest capacitances about 1687 and 998 mAh/g respectively. The Fe$_2$O$_3$/Co$_3$O$_4$ double-helled hollow microcubes acquired the size about 20-30 nm with Coulombic efficiency of 74.4 %. For this composite, the capacity retention is about 30 % after 50 cycles. From the overall analysis of the data on the electrochemical performance of various materials mentioned in Table 1, it is understood that the Li-based materials offered high specific capacitance values with good cyclic ability. This is attributed to the effective ionic motion within Li-based materials. On the other hand, the Co-based material composites also exhibited good electrochemical ability. Thus especially, the Li-ion batteries are well suited for energy storage applications in computers, solar cells, smart mobiles, power tools, grid energy storages, electronic machines, fitness bands, and portable electronic objects. In addition to this, Li-ion battery materials have applications in aerospace. Although the availability of Li-ion battery sources is limited on the earth surface, they have provided applications as the energy storage devices in different fields. Normally, owing to reduced capacity at low temperatures and high temperature cycling of the positive electrode and good energy density and specific energy along with excellent cyclic life of the cathode, the Lithium-ion batteries played a vital role in aviation, which have been used in the systems such as emergency lighting, ELTs, aircraft backup power supplies, cabin doors, UPS (uninterrupted power systems). In drones and fighter jets, these are used in engine start batteries. In addition, these batteries are also used in modern devices such as mobile phones, tablets (USB type-c fast chargers), IoT devices, DSLR cameras, AR/VR devices etc. (see Figure 3). Figure 3 shows the locations of Li-ion batteries in a plane with the suitable usage of it within aerospace [1].

Table1. Various parameters of Li- ion batteries

Sample	Specific capacity (mAh/g)
$LiMn_2O_4$ [5]	154
$LiCoO_2$ [5]	295
$LiNiO_2$ [5]	296
$LiNi_{0.8}Co_{0.15}Al_{0.05}O_2$ [5]	300
$LiNi_{1/3}Mn_{1/3}Co_{1/3}O_2$ [5]	299
$LiFePO_4$ [5]	178
Li_2S [5]	1672
$Li_7Ti_5O_{12}$ [5]	175
$Li_{22}Si_5$ [5]	4200
$Li_{22}Sn_5$ [5]	993
Li_9Al_4 [5]	2235
LiC_6 [5]	372
Lithium metal [5]	3860
3D- $LiFePO_4$/graphene composite [6]	96.7
sepiolite mineral [7]	125
g-C_3N_4 sheet [8]	199.5
$Li_4Ti_5O_{12}$ [9]	161
dual-carbon framework (SnO_2@C/rGO) [10]	844.1
MnO/rGO nanocomposite [11]	2358
Li-ion polysulfide [12]	2430
graphene-modified mesoporous anatase TiO_2 (M-TiO_2-GS) film [13]	205
MnO/C [14]	525
MnO/r-GO [14]	665.5
MnO/C@MRGO [14]	1178
CoO (nano/micro) [14]	1144
CoO/C [14]	755
NiO/MWCNT [14]	800
NiO (110) [14]	700
CuO (nano/micro) [14]	651.6
CuOnanorod [14]	654
CuO/Cu [14]	706
SnO [14]	790
SnO_2-NiO/C [14]	614
SnO_2-NiO/Ag	826
SnO_2/sulfonated graphene [14]	928.5
MoO_2 [14]	1607
Mixed MoO_2 [14]	930.6
WO_3 film [14]	626

WO_3/graphene [14]	656
WO_3/carbon cloth [14]	662
MoO_3 nanoparticle [14]	630
α-MoO_3nanobelt (nb) [14]	1000
MoO_3nb/C [14]	1000
Zn-doped Fe_2O_3 [14]	580
Fe_2O_3nanorods [14]	763
Fe_2O_3nanorod/Ti [14]	562
Fe_2O_3/CNT [14]	375.5
Fe_2O_3/SWCNT [14]	1243
α -Fe_2O_3nanorod [14]	908
Fe_3O_4/N-doped graphene [14]	1187.74
Fe_3O_4/Fe/C [14]	600
Ti-doped Mn_3O_4 [14]	557.8
Mn_3O_4/VGCF [14]	950
CoN film [14]	990
$SiN_{0.79}$ [14]	1500
MnF_2/CNT[14]	203
MgS/C [14]	530
FeP_y (y = 1, 2, 4) [14]	300
FeSnP ($Fe_{51}Sn_{38}P_{11}$) [14]	427
Co_xP (x = 1, 2) [14]	630
CoP/r-GO [14]	960
NiP_2 [14]	1000
P-rich Ni_3P [14]	550
SnSb [14]	700
CeO_2/FeO_3/Mn-rGO composite [15]	420
β-FeOOH nanorods-BC (biomass carbon) [16]	660
Si/C composites [17]	426.1
MoO_3-encapsulated $FeF_{30.33}H_2O$ composites [18]	215.7
$LiNi_{0.6}Co_{0.2}Mn_{0.2}O_2$-based (NCM-622) cathode [19]	123
$LiNi_{0.5}Mn_{1.5}O_4$ [20]	114
$Li_{3/8}Sr_{7/16}Ta_{3/4}Hf_{1/4}O_3$ (LSTH) [21]	142
3D-α-Fe_2O_3 [22]	1157
$Mo_3Nb_2O_{14}$ [23]	329
$(Sn_xCo_{100-x})_{50}C_{50}$ [24]	400
$LiFePO_4$-carbon black films [25]	146.7
$BiVO_4$ microspheres [26]	590
GeP [27]	400
$Li_{1+x}(Ni_{0.8}Co_{0.2})_{1-x}O_2$ [28]	225
$LiCoO_2$ [29]	160
$LiCo_{1/3}Ni_{1/3}Mn_{1/3}O_2$ [30]	140
Pristine -C_3N [31]	1071
hierarchical NiOmicrotubes [32]	770
$Li_{1.123}Mg_{0.010}Ni_{0.2}Co_{0.2}Mn_{0.467}O_2$ [33]	308
Sm_2O_3 coated $Li[Li_{1/6}Mn_{1/2}Ni_{1/6}Co_{1/6}]O_2$ [34]	186

Materials Research Forum LLC
https://doi.org/10.21741/9781644900918-6

N-doped TiO_2/reduced graphene oxide [35]	270
Hierarchical Co_3O_4 [36]	1324
Co_3O_4 nanoboxes [36]	1447
Polyhedral Co_3O_4 [36]	1374
Fusiform Co_3O_4 [36]	1326
Dumbbell-like Co_3O_4 [36]	1719
Co_3O_4 nanorods [36]	1171
Co_3O_4 octahedra [36]	1338
Co_3O_4 nanowires [36]	1027
Co_3O_4 nanotubes [36]	1261
Co_3O_4 nanoparticles /hollow carbon spheres [36]	1249
Co_3O_4 /C nanocapsules [36]	1468
CNF/ Co_3O_4 nanopyramidcoreshell NWs [36]	1173
Co_3O_4 /PCNF [36]	1187
Co_3O_4 /FLGS [36]	1417
Co_3O_4quantumdots/graphene [36]	1630
Co_3O_4/CoO/graphene [36]	1158
CuO/ Co_3O_4 core/shell NW arrays [36]	1633
porous hierarchical Co_3O_4/CuO [36]	1229
Co_3O_4 -α-Fe_2O_3 branched NWs [36]	1534
Fe_2O_3/ Co_3O_4 double-helled hollow microcubes [36]	1687
hierarchicallyFe_2O_3@ Co_3O_4 nanowire array [36]	1587
Co_3O_4 @TiO_2 CSNFs [36]	1034
Nanoporous TiO_2/Co_3O_4 Composite [36]	998
Co_3O_4 /NiO/C core/shell nanowire arrays [36]	1426

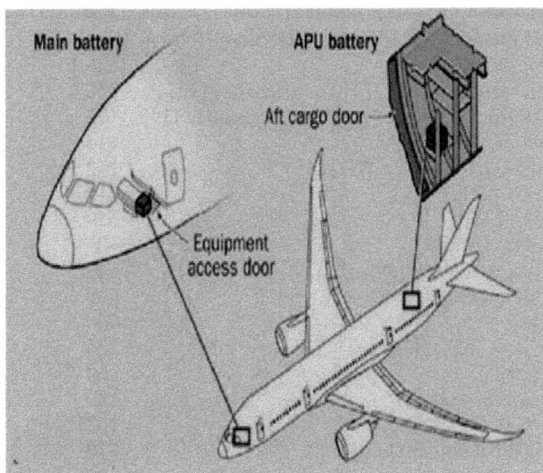

Figure 3:Usage of Li-ion battery in plane.

Conclusions

The $Li_{22}Si_5$, Li-metal, Li-ion polysulfide, Li_9Al_4, and Li_2S recorded the highest specific capacitance values of 4200, 3860, 2430, 2235 and 1672 mAh/g respectively. the Fe_2O_3/Co_3O_4 double-helled hollow microcubes and nanoporous TiO_2/Co_3O_4 composites showed the highest and lowest capacitances about 1687 and 998 mAh/g respectively. This chapter concludes that the Li-based materials offered high specific capacitance values with good cyclic ability owing to the motion of ions within Li-based materials and their composites. Secondly, the Co-based materials have recorded the significant electrochemical activity with high cyclic ability. Therefore, the Li-on batteries have acquired applications in computers, smart mobiles, power tools, grid energy storages, electronic machines, fitness bands, and portable electronic objects as the energy storage devices.

References

[1] Q. Zhaoxiang, G. M. Koenig Jr., Review Article: Flow battery systems with solid electroactive materials, J. Vac. Sci. Technol. B 35 (2017) 040801. https://doi.org/10.1116/1.4983210

[2] E. Ali, Lithium-ion batteries with high rate capabilities, ACS Sustain. Chem. Eng. 5 (2017) 2799–2816. https://doi.org/10.1021/acssuschemeng.7b00046

[3] J.W. Fergus, Ceramic and polymeric solid electrolytes for lithium-ion batteries, J. Power Sources 195 (2010) 4554–4569. https://doi.org/10.1016/j.jpowsour.2010.01.076

[4] J. Munson, Simple calibration circuit maximizes accuracy in Li-Ion battery management systems, Analog Circuit Design Volume 3: Design Note Collection (2015) 419- 420. https://doi.org/10.1016/B978-0-12-800001-4.00197-6

[5] L.H.J. Raijmakers, D.L. Danilov, R.A. Eichel, P.H.L. Notten, A review on various temperature-indication methods for Li-ion batteries, Appl. Energy 240 (2019) 918–945. https://doi.org/10.1016/j.apenergy.2019.02.078

[6] Y. Guan, J. Shen, X. Wei, High-rate performance of a three-dimensional $LiFePO_4$/graphene composite as cathode material for Li-ion batteries, Appl. Surf. Sci. 481 (2019) 1459-1465. https://doi.org/10.1016/j.apsusc.2019.03.213

[7] C. Deng, Y. Jiang, Z. Fan, S. Zhao, D. Ouyang, J. Tan, Y. Ding, Sepiolite-based separator for advanced Li-ion batteries, Appl. Surf. Sci. 484 (2019) 446-452. https://doi.org/10.1016/j.apsusc.2019.04.141

[8] J. Zhang, G. Liu, H. Hu, L. Wu, Q. Wang, X. Xin,P. Lu, Graphene-like carbon-nitrogen materials as anode materials for Li-ion and mg-ion batteries, Appl. Surf. Sci.487 (2019) 1026-1032. https://doi.org/10.1016/j.apsusc.2019.05.155

[9] F. Khan, M. Oh, &J. H. Kim, N-functionalized graphene quantum dots: charge transporting layer for high-rate and durable $Li_4Ti_5O_{12}$–based Li-ion battery, Chem. Eng. J. 369 (2019) 1024- 1033. https://doi.org/10.1016/j.cej.2019.03.161

[10] H. Li, B. Zhang, Q. Zhou, J. Zhang, W. Yu, Z. Ding, A.M. Tsiamtsouri, J. Zheng and H. Tong, Dual-carbon confined SnO_2 as ultralong-life anode for Li-ion batteries, Ceram. Int. 45 (2019) 7830-7838. https://doi.org/10.1016/j.ceramint.2019.01.090

[11] H. Raj, A. Sil, N.V. Pulagara, MnO anchored reduced graphene oxide nanocomposite for high energy applications of Li-ion batteries: The insight of charge-discharge process, Ceram. Int. 45 (2019) 14829–14841. https://doi.org/10.1016/j.ceramint.2019.04.214

[12] A. Sawas, G.S. Babu, N.K. Thangavel, L.M.R. Arava, Electrocatalysis driven high energy density Li-Ion polysulfide battery, Electrochim. Acta 307 (2019) 253- 259. https://doi.org/10.1016/j.electacta.2019.03.191

[13] H. Luo, C. Xu, B. Wang, F. Jin, L. Wang, T. Liu,D. Wang, Highly conductive graphene-modified TiO_2 hierarchical film electrode for flexible Li-ion battery anode,Electrochim.Acta 313 (2019) 10-19. https://doi.org/10.1016/j.electacta.2019.05.018

[14] M. D. Bhatt, J. Y. Lee, High capacity conversion anodes in Li-ion batteries: A review, Int. J. HydrogenEnergy44 (2019) 10852-10905. https://doi.org/10.1016/j.ijhydene.2019.02.015

[15] K.O. Ogunniran, G. Murugadoss, R. Thangamuthu, S.T. Nishanthi, Nanostructured $CeO_2/FeO_3/Mn$-rGO composite as anode material in Li-ion battery, J. AlloysCompd. 786 (2019) 873–883. https://doi.org/10.1016/j.jallcom.2019.02.024

[16] C. Wang, X. Yang, M. Zheng, Y. Xu, Synthesis of β-FeOOH nanorods adhered to pine-biomass carbon as a low-cost anode material for Li-ion batteries, J. AlloysCompd. 794(2019)569-575. https://doi.org/10.1016/j.jallcom.2019.04.074

[17] K. Wang, B. Xue, Y. Tan, J. Sun, Q. Li, S. Shi, P. Li, Recycling of micron-sized Sipowder waste from diamond wire cutting and its application in Li-ion battery anodes, J. Clean.Prod. 239 (2019) 117997. https://doi.org/10.1016/j.jclepro.2019.117997

[18] X. Zhou, J. Ding, J. Tang, J. Yang, H. Wang, M. Jia, Tailored MoO_3-encapsulated $FeF_{30.33}H_2O$ composites as high performance cathodes for Li-ion batteries, J.Electroanal. Chem.847 (2019) 113227. https://doi.org/10.1016/j.jelechem.2019.113227

[19]   C. Heubner, A. Nickol, J. Seeba, S. Reuber, N. Junker, M. Wolter, M. Schneider, A. Michaelis, Understanding thickness and porosity effects on the electrochemical lperformance of $LiNi_{0.6}Co_{0.2}Mn_{0.2}O_2$-based cathodes for high energy Li-ion batteries, J. Power Sources 419 (2019) 119–126. https://doi.org/10.1016/j.jpowsour.2019.02.060

[20]   T. Kozawa, Lattice deformation of $LiNi_{0.5}Mn_{1.5}O_4$ spinel cathode for Li-ion batteries by ball milling, J. Power Sources 419 (2019) 52–57. https://doi.org/10.1016/j.jpowsour.2019.02.063

[21]   B. Huang, S. Zhong, J. Luo, Z. Huang,C. Wang, Highly dense perovskite electrolyte with a high Li^+ conductivity for Li–ion batteries, J. Power Sources 429 (2019) 75–79. https://doi.org/10.1016/j.jpowsour.2019.04.117

[22]   J. Park, H. Yoo, J. Choi, 3D ant-nest network of α-Fe_2O_3 on stainless steel for all-in-one anode for Li-ion battery, J. Power Sources 431(2019) 25–30. https://doi.org/10.1016/j.jpowsour.2019.05.054

[23]   P.M. Ette, D. B. Babu, M. L. Roy, K. Ramesha, $Mo_3Nb_2O_{14}$: A high-rate intercalation electrode material for Li-ion batteries with liquid and garnet based hybrid solid electrolytes, J. Power Sources 436 (2019) 226850. https://doi.org/10.1016/j.jpowsour.2019.226850

[24]   A.S. Kumar, M. Srinivas, A.V.P. Kiran, L. Neelakantan, Structural and electrochemical properties of $(Sn_xCo_{100-x})_{50}C_{5c}$ anodes for Li-ion batteries, Mater. Chem. Phys. 236 (2019) 121782. https://doi.org/10.1016/j.matchemphys.2019.121782

[25]   X. Michaud, K. Shi, I. Zhitomirsky, Electrophoretic deposition of $LiFePO_4$ for Li-ionbatteries, Mater. Lett.241(2019) 10-13. https://doi.org/10.1016/j.matlet.2019.01.032

[26]   D.R. Patil, S.D. Jadhav, A. Mungale, A.S. Kalekar, D.P. Dubal, Fractal granular $BiVO_4$ Microspheres as high performance anode material for Li-ion battery, Mater. Lett. 252 (2019)235-238. https://doi.org/10.1016/j.matlet.2019.05.142

[27]   W. Li, X. Li, J. Yu, J. Liao, B. Zhao, L. Huang, A. Abdelhafiz, H. Zhang, J.H. Wang, Z. Guo, M. Liu, A self-healing layered GeP anode for high-performance Li-ionbatteries enabled by low formation energy, Nano Energy 61 (2019)594-603. https://doi.org/10.1016/j.nanoen.2019.04.080

[28]   R. Wang, G. Qian, T. Liu, M. Li, J. Liu, B. Zhang, F. Pan, Tuning Li-enrichment in high-Ni layered oxide cathodes to optimize electrochemical performance for Li-ion battery, Nano Energy62 (2019) 709-717. https://doi.org/10.1016/j.nanoen.2019.05.089

[29]   R.A. Adams, B. Li, J. Kazmi, T.E. Adams, V. Tomar, V.G. Pol, Dynamicimpact of $LiCoO_2$ electrodes for Li-ion battery aging evaluation, Electrochim.Acta292 (2018)586-593. https://doi.org/10.1016/j.electacta.2018.08.101

[30]   S. Ahn, H.S. Kim, S. Yang, J.Y. Do, B.H. Kim, K. Kim, Thermal stability and performance studies of $LiCo_{1/3}Ni_{1/3}Mn_{1/3}O_2$ with phosphazene additives for Li-ion batteries, J. Electroceram. 23 (2008) 289–294. https://doi.org/10.1007/s10832-008-9437-y

[31]   G.C. Guo, R.Z. Wang, B.M. Ming, C. Wang, S.W. Luo, C. Lai, M. Zhang,Trap effects on vacancy defect of C_3N as anode material in Li-ion battery, Appl. Surf. Sci. 475 (2019) 102–108. https://doi.org/10.1016/j.apsusc.2018.12.275

[32]   Y. Hong, J. Yang, J. Xu, W. M. Choi, Template-free synthesis of hierarchical NiO microtubes as high performance anode materials for Li-ion batteries, Curr. Appl. Phys. 19 (2019) 715–720. https://doi.org/10.1016/j.cap.2019.03.019

[33]   Y. Jin, Y. Xu, F. Ren,P. Ren,Mg-doped $Li_{1.133}Ni_{0.2}Co_{0.2}Mn_{0.467}O_2$ in Li site as high-performance cathode material for Li-ion batteries, Solid State Ionics 336 (2019) 87–94. https://doi.org/10.1016/j.ssi.2019.03.020

[34]   G. Y. Kim, Y. J. Park, Enhanced electrochemical and thermal properties of Sm_2O_3 coated $Li[Li_{1/6}Mn_{1/2}Ni_{1/6}Co_{1/6}]O_2$ for Li-ion batteries, J. Electroceram. 31 (2013) 199–203. https://doi.org/10.1007/s10832-013-9806-z

[35]   J. Li, J. Huang, J. Li, L. Cao, H. Qi, Y. Cheng,N-doped TiO_2/rGO hybrids as superior Li-ion battery anodes with enhanced Li-ions storage capacity, J. Alloys Compd. 784 (2019) 165–172. https://doi.org/10.1016/j.jallcom.2019.01.061

[36]   Y. Shi, X. Pan, B. Li, M. Zhao, H. Pang, Co_3O_4 and its composites for high-performance Li-ion batteries, Chem. Eng. J.343 (2018) 427 – 446. https://doi.org/10.1016/j.cej.2018.03.024

Lithium-ion Batteries - Materials and Applications
Materials Research Foundations **80** (2020) 161-202

Materials Research Forum LLC
https://doi.org/10.21741/9781644900918-7

Chapter 7

Silicon Materials for Lithium-ion Battery Applications

Sarigamala Karthik Kiran[1], Martha Ramesh[2], Shobha Shukla[2] and Sumit Saxena[2]

[1] Centre for Research in Nanotechnology and Science, Indian Institute of Technology Bombay, Mumbai, MH, India - 400076

[2] Nanostructures Engineering and Modeling Laboratory, Department of Metallurgical Engineering and Materials Science, Indian Institute of Technology Bombay, Mumbai, MH, India - 400076

Abstract

Silicon, a hard brittle crystalline solid is tetravalent metalloid and prominently well known in the semiconductor community. Although silicon has been a material of choice for energy generation in silicon solar cells, its high theoretical lithium storage capacity makes it one of the most promising anode materials for development of high performance Li-ion batteries. Unfortunately, silicon exhibits large volume expansion leading to severe problems associated with structural integrity of the electrode and capacity retention. Several silicon nanostructures have been explored to mitigate shortfalls of using bulk silicon as electrode material. In this chapter, we discuss various promising designs, utilizing various nanostructures as a possibility to mitigate the issues related to the use of silicon in Li-ion batteries. The formation of axial heterojunctions, and core/shell nanostructures is discussed. Processes such as etching based metal assisted electrochemical, co-precipitation, magnesiothermic reduction, and chemical vapour deposition techniques are briefly discussed. The structural, electrical and electro-chemical properties of different nanostructures grown by these methods are also summarized.

Keywords

Silicon Anode, Li-ion Battery, Nanostructures, Core Shells, Capacity, Stability

Contents

1.   Introduction

We are largely dependent on use of fossil fuel to meet our current energy requirements. The combustion of these fuels for transport, industrial and domestic purposes has not only affected human health but also adversely affected the entire ecosystem [1]. Additionally, it takes millions of years and extreme conditions to convert fossilized matter into fuel inside the earth. This process of generation of fossil fuel finds no match to the rate of extraction, which is anticipated to render the earth depleted of fossil fuel in decades to come [2]. Unfortunately, no technology is available today to replace the use of fossil fuels entirely. It is believed that technologies like hydrogen fuel, water, wind, and solar may collectively overcome the dependency on fossil fuel [3,4]. Most alternative forms of energy generation are focussed on generation of electricity. It is not only challenging to store but also requires significant infrastructure to transmit. Some interesting initiatives using solar power have been explored to make high energy density fuels from renewable resources [5]. However, these may be more expensive in terms of energy generation/storage and require huge infrastructure to transport and store them safely. The problem with such renewable power sources is that they might not be produced at the time when needed, especially during periods of peak energy demand [6]. Further, the environmental factors also play an important role in generation of energy using natural resources [7]. Thus in order to provide steady power supply using renewable energy sources, energy storage seems the best way forward.

Lithium-ion Batteries - Materials and Applications   Materials Research Forum LLC
Materials Research Foundations **80** (2020) 161-202   https://doi.org/10.21741/9781644900918-7

While several strategies have been explored to store energy in various forms, storage of energy electrochemically in form of batteries and supercapacitors forms one of the most explored and preferred energy storage technique [8]. The term battery was coined by Benjamin Franklin for multiple Leyden jars, in analogy to battery of cannons. However, the history of batteries is believed to be more than 2000 years old when prehistoric batteries in form of Parthian batteries were excavated near Baghdad. However, first modern batteries are supposed to have made their mark in mid of 17th century with advent of Leyden jars. Although early batteries showed great promise for experimental purposes, they could not find large scale applications due to voltage fluctuation and their inability to provide large current for long durations [9]. Lead-acid battery invented by Gaston Plante in 1859 is one of the oldest known rechargeable battery. Although the lead-acid battery had a very low energy-to-weight ratio and a low energy-to-volume ratio, it could provide huge surge current requirements. This made them attractive for automobile applications where high current is needed for automobile starter motors [10]. While several discoveries were made over years by many innovators like Galvini, Volta, Davy, Cruickshank and Daniell. Waldemar Jungner in 1899 provided a breakthrough by developing Ni-Cd cells which changed the perception of storage of energy by providing portable energy sources [11]. For many years Ni-Cd batteries remained the flagship for portable energy devices. However, careless disposal of these batteries resulted in environmental hazards which restricted its usage since 2006 by European Union. While the success story of Ni-Cd batteries was unfolding, yet another initiative made its headway lead by an American physicist and Nobel Laureate, Professor John Goodenough which could compete with the Ni-Cd batteries in making available portable energy. These batteries worked on the principle of migration of Li$^+$ ion between the electrodes. It was first commercialized by SONY in 1991.

1.1   Overview on lithium battery technology

Lithium forms the element of choice for battery applications due to the following reasons [12,13]. (a) Its ability to readily lose its outermost electron makes it highly reactive resulting in easy flow of current in a battery application. (b) It has lower reduction potential as compared to any other element in the periodic table. (c) Last but not the least, its availability in significant amounts in natural resources. Lithium metal batteries refer to battery systems comprising of metallic lithium as anode with all possible materials combination for cathode and an electrolyte [14]. Although they offer high charge density, high operating voltages (1.5V to 3.7V) and longer life, the downside is higher cost per unit cell. One of the earliest developments in this area was made by Whittingham in 1970's while working for Exxon. He used titanium (IV) sulfide and lithium metal as electrodes for making a lithium battery [15]. This however could not see the light of the

Materials Research Forum LLC

https://doi.org/10.21741/9781644900918-7

day as it not only required controlled environment for fabrication but also it was economically not feasible. Further metallic lithium being a highly reactive element, ignites spontaneously on exposure to atmosphere due to spontaneous reaction with water and oxygen. This raised serious safety concerns on operation and use of such battery as a power source [16]. The inability to use metallic lithium due to its stability issues resulted in exploration of non-metallic lithium batteries using lithium ions. These batteries are commonly used as rechargeable batteries till today due to their high energy densities and economic feasibility [17]. They also provide wide operating voltages of about ~3.6 volts making a single cell sufficient to power most portable devices. These characteristics of Li ion batteries have raised the market demand and also paved a new path to many innovations in the areas of materials research, engineering and technology. It has finally placed this technology in an esteem position as a battery of the future [18].

The energy-density of Li-ion battery primarily depends on specific capacity of the electrode materials. The energy density and longer cycling life of these batteries can be enhanced by increasing the specific capacity of cathode and anode materials [19-21]. The specific capacity of cathode has been improved to 160 and even to 200 mAhg^{-1} with new material systems such as $LiMn_{1-x}M_xO_2$ [22], $Li[Ni_xCo_{(1-2x)}Mn_x]O_2$ [23], defective Li–Mn–O spinel structures [24], olivine $LiFePO_4$ [25], when compared to the traditional $LiCoO_2$ material (135 mA.h.g^{-1}).

Various modification techniques, such as coating or compositing, elemental doping and combination with mesoporous materials have been explored for preparing nanostructured cathode materials [26]. These materials have both positive and negative effects on the electro-chemical properties of the cell. Some of the drawbacks include poor cycling stability, low conductivity and cationic disorder. Low electrical conductivity is one of the major problems that lead to poor electrochemical performance of cathode materials. The use of composite or coated materials with high conductivities serves two purposes,

1) Improves the overall conductivity of the entire composite material

2) Prevents the reunion of the particles, which facilitate short reaction pathways for insertion or extraction of lithium ions [27].

However, the preparation process of nanocomposites is often critical. It requires significant time to achieve a good dispersion, controllability and good yield of nanoparticles in the composite. However, it becomes an expensive and low cost-competitive initiative [28].

Several multiple redox mechanisms have been identified in transition metal oxides for charge storage which is termed as conversion reaction [29]. Multiple charge carrier pairs are produced during the redox process of nanocrystals. Subsequently it results in enhanced energy conversion efficiency (electrical to chemical energy and vice versa). Such electrode materials include iron oxides, which are relatively abundant and easy to prepare. The lithium storage in the negative electrodes is also based on the reversible redox reactions that produce high specific capacities. However, the high conversion efficiency and reversibility is only possible in nanoscale particles. Hence, engineering novel nanostructures would probably be an effective strategy to produce high-performance LIBs based on the anodes (Fig. 1).

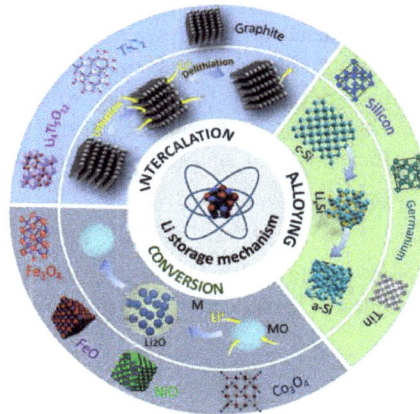

Figure 1. Different Li storage mechanisms of anode materials. (Adopted with permission from ref [30]. © 2017 Elsevier)

Lithium ion batteries involve three charge transfer reaction mechanisms:

a) Intercalation/de-intercalation:

This process refers to the reversible insertion/extraction of lithium to/from a solid host network. These solid networks may have different crystal structures such as layered, spinel or olivine. The intercalation processes do not usually result in a huge change in volume. This accounts for the stability of spinel $Li_4Ti_5O_{12}$ and graphite as electrodes during the charge/discharge cycles. Additionally, these provide desirable properties such as low cost, abundant availability, low de-lithiation potential vs. Li, high electronic conductivity, and good ion diffusivity. [31] However, intercalation processes are usually not very efficient in terms of guest ions compared to host ions. Graphite, for example, can store one Li-ion for every 6 carbon atoms and results in relatively poor theoretical gravimetric capacity (372 mAh g^{-1}) [32]. This acts a bottleneck in meeting the high energy demands for applications in automobile industry. This has led to exploration of a large number of materials to achieve high energy densities, improved storage capacity, and improved cycle characteristics.

b) Conversion reaction mechanism:

Conversion materials store Li ions according to the following mechanism:

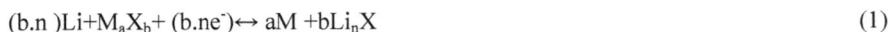

$$(b.n\ )Li + M_aX_b + (b.ne^-) \leftrightarrow aM + bLi_nX \tag{1}$$

Here, M represents a transition metal and X represents an anion, which is usually oxygen in the case of anodes. This class of materials tends to have a lot of beneficial properties, in particular, they show a high capacity (e.g. MnO2 with 755mAh g^{-1} [33]), are inexpensive and environment-friendly. However, they also show significant drawbacks such as low Coulombic efficiency (CE), unstable solid electrolyte interface (SEI) and poor cycling stability [34].

c) Alloying/de-alloying:

This forms the third process due to which Li storage can take place and the reaction proceeds according to the following general mechanism,

$$M + yLi \leftrightarrow Li_yM \tag{2}$$

where M corresponds to either Si, Ge, Sn or Pb. Storing Li in materials by alloying shows very high (theoretical) capacities, as well as high energy densities and safe operations, but they show severe problems when it comes to cyclic stability and capacity retention.

1.2   Silicon as anode for lithium batteries:

Amongst all anode materials, silicon (Si) has attracted significant attention for Li-ion batteries. Si is a well-known and abundantly available semiconductor. It finds applications in solar cells, microprocessors, and liquid crystal displays. Additionally, Si has superior properties, such as high theoretical specific capacity of 4,200 mAhg^{-1} and high volumetric capacity of 9,786 mAh cm^{-3} [35, 36]. It offers storage capacity which is 10 times as compared to carbon based materials. Si additionally shows some beneficial intrinsic properties, such as low delithiation potential vs. Li/Li$^+$ (0.4V) compared to most other anode materials, which results in a high working voltage and energy density [37]. Since the potential is higher than that of graphite (0.05V vs. Li/Li$^+$), it is less likely to form Li dendrites at high current densities. The dendrites often lead to short circuit and ignition or explosions [38] in the cell.

In a conventional Si-Li cell, shown in Fig. 2 (a-b), the silicon electrode sits on the top of a separator and is patterned to provide access points to contact the lithium cathode. These access point allows electrons to transfer from negative to the positive terminal as shown in Fig. 2a, completing the circuit. The porous membrane separator is soaked in a non-aqueous liquid electrolyte. The Li^+ ions de-intercalate from the anode during the discharging cycle, pass through the electrolyte and are inserted into the cathode. The electrons flow from the anode to the cathode through the external circuit and powers the load attached across the cell. A reverse process occurs during the charging cycle on applying a potential across the battery. Lithium ion batteries work on the principle of reversible shuttling of Li^+ ions. The difference of electrochemical properties of the electrodes and their inherent properties play a crucial role in determining overall performance of the batteries.

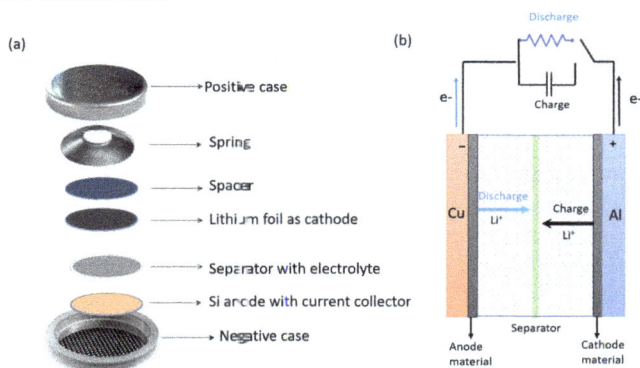

Figure 2. (a) Assembled Si-Li coin cell schematic and (b) Charge-discharge mechanism schematic.

Silicon can either be amorphous (a-Si), poly-crystalline (p-Si) or crystalline (c-Si), with the latter showing the highest electronic conductivity of all the three. c-Si undergoes a phase transition upon first lithiation changing from a crystalline to an amorphous state, and remains amorphous during entire cycling. A combination of focused ion beam time-of-flight secondary ion-mass-spectrometry and auger electron spectroscopy was used by Bordes et al. to investigate Si particles during the first lithiation process [41]. The first 5% of Li form a solid-electrolyte interface (SEI), followed by a two-phase lithiation, characterized by a sharp edge between the alloy shell and the pure Si core. Fast diffusion pathways for the Li penetrate the pure Si core due to sub-grain boundary pathways as shown in Fig. 3. After 70% lithiation, the core-shell structure disappears and the particle starts to disintegrate [41].

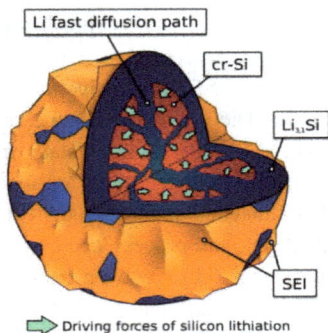

Figure 3. Schematic of alloying mechanisms of a Si particle during first lithiation. A two phase mechanism with a sharp boarder between core (pure Si) and shell (alloy) and Li diffusion pathways penetrating the core has been observed. (Adopted with permission from ref [41]. © 2016 ACS)

A clear boundary between the lithiated and pure Si phase is observed in transmission electron micrographs as shown in Fig. 4 suggesting anisotropic lithiation in c-Si along with facet formation during the first lithiation. The c-Si remains amorphous after full lithiation and further cycling occurs in an isotropic, single phase mechanism (Fig. 4) [42].

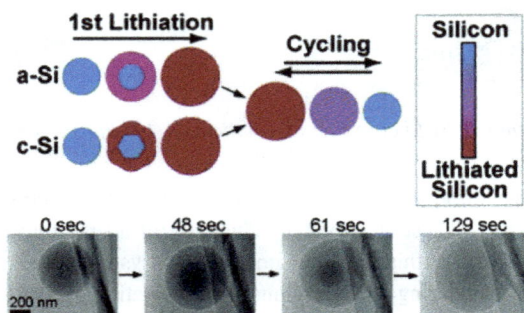

Figure 4. TEM images of a Si nanoparticle during the first lithiation, showing a two phase mechanism. The schematic shows the isotropic lithiation of a-Si compared with the anisotropic lithiation with facet formation of c-Si during the first cycle. After full lithiation, c-Si remains in an amorphous state. During further cycling, lithiation is isotropic and a one-phase mechanism. (Adopted with permission from ref [39]. © 2017 RSC)

The ability of Si to store large amounts of Li results in huge volumetric changes up to 420% during lithiation/de-lithiation [43]. This results in strong capacity fading and cell failure, mainly due to the following three effects as suggested in Fig. 5. Pulverization and cracking of the anode material occurs due to stress induced by the volume change upon lithiation/delithiation. This leads to loss of electric contact of the active material with the conducting framework or the current collector resulting in capacity fading upon cycling [44]. Delamination and morphology changes are a result of structural change upon cycling. During lithiation, the Si particles or structures expand and impinge on each other, whereas during delithiation they contract leading to loss of conductive contact [45]. On a larger scale, the whole electrode expands and contracts, which can cause cell failure [43]. Unstable Solid-Electrolyte-Interface (SEI) formation can be attributed to irreversible Li-ion consumption due to their reduction on the electrode surface and leading to building up of SEI. This layer is conductive for Li and covers the active material, which further prevents the SEI build up and results in a smooth and thin SEI layer [46].

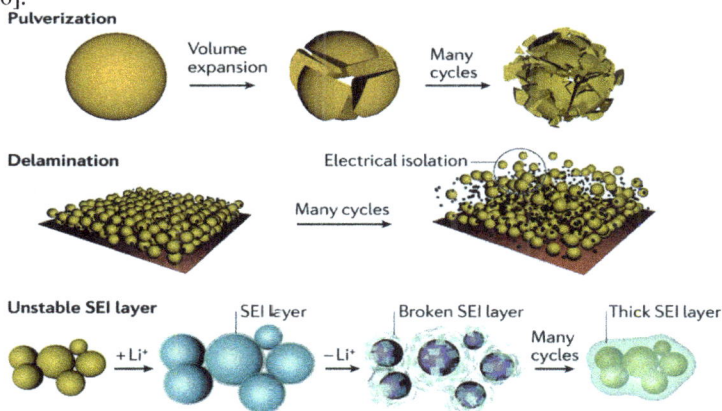

Figure 5. Schematic showing three main effects i.e. pulverization, delamination and unstable SEI formation that contribute to capacity fading and cell failure (Adopted with permission from ref. [39]. © 2017 RSC)

However, due to the large expansion of Si during lithiation, the SEI cracks open and expose fresh Si. Subsequently, electrolyte gets consumed and an inhomogeneous SEI with high resistance is formed. However, a thin, smooth and stable SEI is crucial to obtain little loss in capacity during cycling [47].

In order to address these issues and enhance the performance of anode materials various strategies have been explored. Fig.6 shows an overview of the most common ones. The main goals are to compensate or accommodate the huge volume change of Si during cycling, facilitate electrochemical reactions by providing fast and short lithium ion diffusion pathways, maintain a high surface to volume ratio as well as good electronic connectivity with the current collector and ensure a smooth, thin and stable SEI formation [39].

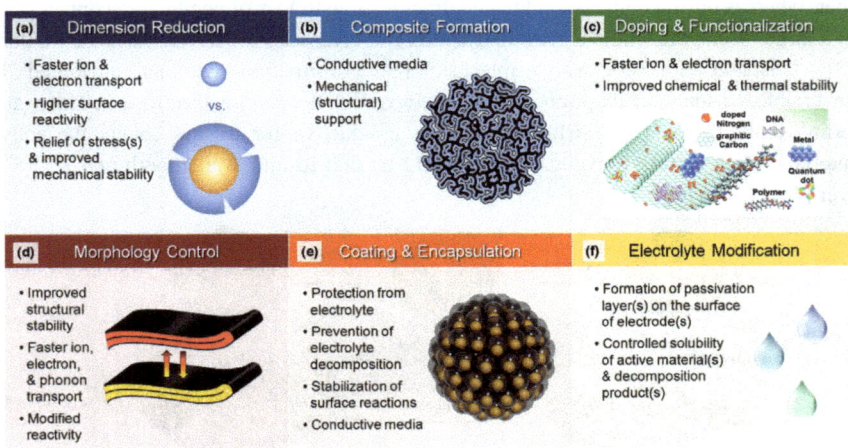

Figure 6. Overview of different approaches to enhance anode performance and counteract the problems associated with Si as an anode material. Most of them incorporate in one way or another field of nanoscience. (Adopted with permission from ref [31. © 2015 Elsevier)

As lithium atoms are added to the Si lattice, lattice expansion occurs and alloy formation starts to set in. Once the alloy is formed completely, the volume of the new fully lithiated alloy increases by four fold of the original Si lattice. This may trigger development of cracks and failing of the electrode results in the loss of electrical contact. In order to mitigate these issues, different approaches have been explored to achieve better electrochemical performance by using novel silicon nanostructures in the form of nano-/micro-Si particles, thin films, nanotubes, nanowires, bulk porous morphologies and yolk-shell silicon nanostructures as shown in Fig. 7. Scanning electronic microscopy along

Materials Research Forum LLC

https://doi.org/10.21741/9781644900918-7

with theoretical calculations have been employed to investigate these processes [48-52]. Silicon nanocomposites are being explored to improve the mechanical stability of the electrodes [53-56].

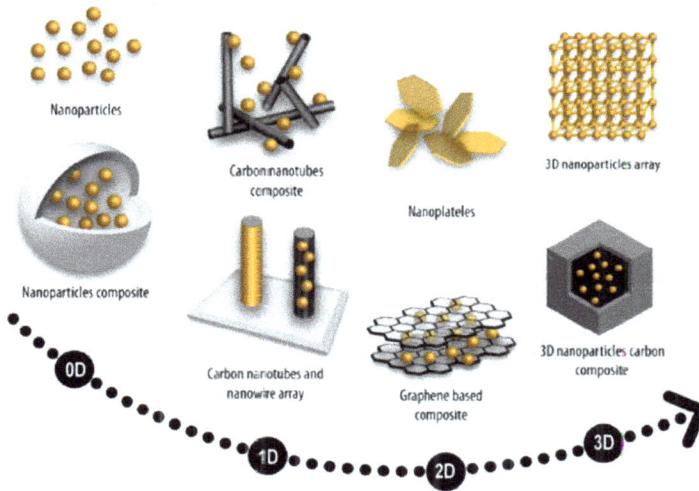

Figure 7. Most common nano-structures found in novel anode designs. (Adopted with permission from ref [57]. © 2017 Springer)

Anodes based on pure Si nanoparticles do exhibit a large surface area, leading to unwanted side reactions and extensive SEI formation. Often, these effects are counteracted by implanting the nanoparticles into a stabilising and/or conducting matrix or grid [58]. This however may result in electrochemical aggregation as the nanoparticles tend to accumulate upon many cycles of expansion and contraction and disconnect from the current collector. 1D nanorods, nanowires and nanotubes provide good conductivity due to the 1D transport channel along them, and hence they are usually directly grown on the current collector [58]. Furthermore, the space between the nanorods can accommodate volumetric change, all of which leads to a high cycling stability as well as a high reversible capacity [59]. Nanotubes additionally provide accessible channels for electrolyte transport and short lithium ion diffusion pathways, as well as internal space for volume expansion. The high specific surface of 1D structures unfortunately leads to more irreversible reactions and thus to a higher SEI build up [58]. 2D thin films and nanoplates generally show a high capacity retention upon cycling, but they tend to crack upon the first lithiation and hence show strong capacity fading during the first cycle. It has been observed in the case of thin films that their decreasing thickness increases the reversible capacity and cycling stability [39]. Si nanosheets exhibit very short lithium diffusion pathways, little volume expansion and are highly compatible with other materials [58]. 3D Si-based porous structures and networks can accommodate volumetric changes due to their porous character and thus maintain structural integrity while having a much higher tap density. Thus theoretically a much higher volumetric capacity, than nanostructures of lower dimensionality is achieved. 3D structures also provide fast lithium diffusion through their pores, but similar to 0D and 1D approaches, composites usually show better performance than pure Si structures [39, 58].

1.2.1 0D nanostructures

The term 0D nanostructures is generally used to represent nanocrystals, nanoparticles, nanodots, or quantum dots. These can be synthesized using different techniques such as molecular beam epitaxy (MBE), sputtering, laser ablation, and chemical methods [60–62]. A variety of nanostructured Si or Si-C materials have been explored as anode materials [63-67]. Si/Carbon core-shell composite synthesized shows a capacity retention of 74% after 1000 cycles [68]. However, synthesis of hierarchically-structured Si-anodes are often elaborate and involve complicated treatment like pre-coating or selective etching [69-70]. These processes may require use of hazardous chemicals like HF and severe conditions like high temperature [71,72]. These complicated synthesis procedures add to the production costs and undermines the commercial attractiveness of silicon [73,74]. On the other hand, carbon-based hosts have poor mechanical stability. This leads to fracture, thereby resulting in continued exposure of fresh surface of the Si core to the electrolyte, which causes sustained capacity loss after numerous cycles [75]. Although the introduction of a protective metal oxide layer (e.g., SiO_x, TiO_2, TiO_2-xF_x and Al_2O_3) is expected to increase the interfacial and structural stability of Si-based composites in the low potential range, these coating materials usually show poor electrode kinetics [76-80]. Therefore, it is important to design a Si-based composite, functionalized with a coating layer that is conducting (both ionically and electrically). In contrast to other techniques, co-precipitation is one of the most convenient techniques for either preparation of nanoparticles or the incorporation of trace elements into nanospheres with a narrow size distribution, during solid solution formation and recrystallization. In addition, the co-precipitation technique does not require costly equipment, stringent reaction conditions, or complex procedures [81]. In some cases, crystalline nanoparticles can be obtained directly by co-precipitation without calcination or post annealing process. Surface defects form a matter of concern for battery applications. Surface passivation of nanocrystals occurs during wet chemical synthesis, this results in suppression of interfacial defects and carrier recombination at the surface [82].

Another approach which has been explored in the recent past has been the use of Metal-organic frameworks (MOFs) [83, 84]. Some of the low cost potential candidates include Prussian blue and its analogues (PBAs) primarily due to possibility of achieving highly reversible insertion and extraction of ions. These have been widely used as water colours and catalysts [85-88]. KMFe(CN)$_6$ compounds (where M = Mn, Fe, Co, Ni and Zn) when used as cathodes in silicon batteries do not show capacity loss after 30 cycles [89]. MnFe(CN)$_6$ demonstrated specific capacity of 295.7 mAh g^{-1} after 200 cycles as the anode for lithium ion batteries [90]. However, limited success with lower capacity of PBA anodes [91-92] lead to exploration of core-shell Si-based nanocomposites in lithium ion batteries for confining the volumetric change and enhancing the electronic conductivity of Si anode.

The schematic for synthesis of PBAs@Si nanocomposites using co-precipitation method at room temperature is shown in Fig.8. Na$_4$Fe(CN)$_6$ and MnCl$_2$ are used as ion source in the presence of Si NPs suspension. The [Fe(CN)$_6$]$^{4-}$ anions chemisorb onto the surface of silicon nanoparticles forming SiO$_2$ core. The silicon nanoparticles act as nucleation sites resulting in formation of PBAs on the surface of silicon nanoparticles. Subsequent reaction with [Fe(CN)$_6$]$^{4-}$ results in the further growth of PBAs on the PBAs@Si crystal. The color of the sample changes gradually from pale yellow to yellow-green.

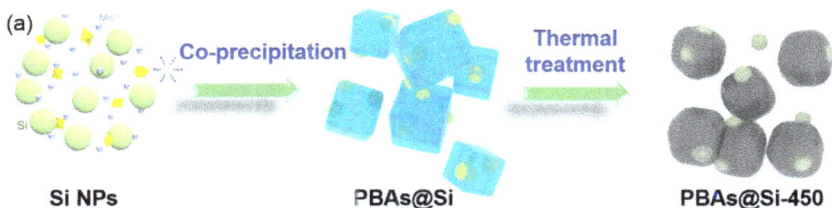

Figure 8. Schematic for the synthesis of PBAs@Si-450 nanocomposite. (Adopted with permission from ref [93].© 2018 ACS)

1.2.2 1D nanostructures

Nanowires, nanotubes, and nanopores are commonly referred to as 1D nanostructures due to large aspect ratio. These have been grown and fabricated using variety techniques such as chemical vapour deposition, molecular beam epitaxy, laser ablation and metal-catalysed chemical etching (MACE) to name a few [94-101]. MACE is an interesting technique to grow Si nanostructures [102-104]. In this process, the noble metals spots are deposited on silicon surface using techniques like thermal evaporation [102-105,106], sputtering [107], electron beam evaporation [108], electroless deposition [109] and spin coating [110]. The substrate in MACE do not act as catalyst but as electron donors, so that it can be etched to assist in the formation of nanostructured patterns [111].

The MACE mechanism for the preparation of Si NPs and Si NWs is briefly described below. The Ag^+ ions deposited on the silicon surface capture electrons from the valence band of Si and forms metallic Ag nuclei. The electron transfer from silicon to silver ions occurs due to the higher electro-negativity of the silver ions. The silver nuclei so formed assists in preferential reduction of silver ions resulting in deposition of more metallic silver and growth of silver nuclei. However, during the process excessive local oxidation occurs beneath the silver particles resulting in formation of SiO_2. The underlying SiO_2 beneath the metallic silver nuclei dissolves leading to formation of pits and subsequently deepening occurs resulting in formation of straight, non-interacting cylindrical pores orthogonal to the Si surface as seen in Fig. 9. Further dissolution of the thin pore walls the adjacent pores merge leading to formation of large quantities of silicon nanowires on the silicon surface. In this process, the morphology of the produced silicon nanowires depends crucially on the etching parameters such as solution concentration, temperature and metal species in the HF solution.

The following reactions are involved [102-104]:

$$HF \rightarrow F^- + H^+$$

$$Si + 6F \rightarrow SiF_6^{2-} + 4e^-$$

$$AgNO_3 \rightarrow Ag^+ + NO_3^-$$

$$Ag^+ + e^- \rightarrow Ag^0(s)$$

Lithium-ion Batteries - Materials and Applications
Materials Research Foundations **80** (2020) 161-202

Materials Research Forum LLC
https://doi.org/10.21741/9781644900918-7

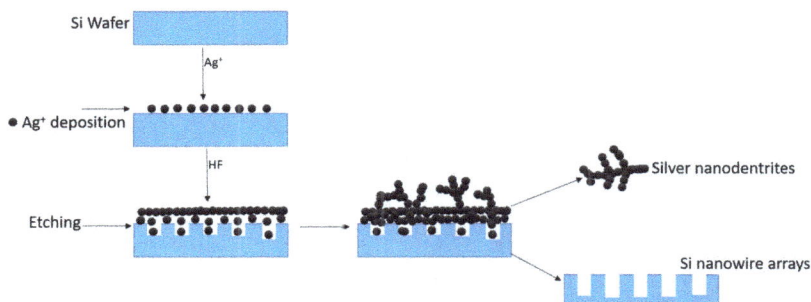

Figure 9. Schematic of MACE for the preparation of Si NPs and Si NWs.

1.2.3 2D nanostructures

2D nanostructures are commonly known as thin films, nanoplatelets and nanosheets. The 2D nanostructures have typically been synthesized using chemical decomposition of silanes [112, 113] or sputtering [114, 115]. Template based method to synthesize silicon nanosheets have been explored recently. However, the process leads to small sheets with low yield and have poor electrochemical properties [116]. In another approach the silicon nanosheets were produced using sand and the resulting silicon sheets were broken into smaller nanosheets. The sheets so produced showed formation of stacked structures resulting in detrimental effects such as lower Coulombic efficiencies and drastic capacity decays in performance of the cell [117]. Natural clays containing silicates with layered structures have also been explored to produce 2D silicon nanosheets by exfoliating then using molten salts [118, 119]. Unfortunately, multiple elements present in the clay makes it difficult to remove these other elements.

Hydrolyzed tetraethyl orthosilicate (TEOS) as the silica source with NaCl particles as templates have also been used to synthesize 2D silicon nanosheets. The schematic is shown in Fig. 10. This process involves in situ synthesis of SiO_2 nanosheets accompanied with reduction of Mg. SiO_2 nanosheets are formed using the surface of NaCl particles as template and the TEOS as the silicon precursor.

Lithium-ion Batteries - Materials and Applications   Materials Research Forum LLC
Materials Research Foundations 80 (2020) 161-202   https://doi.org/10.21741/9781644900918-7

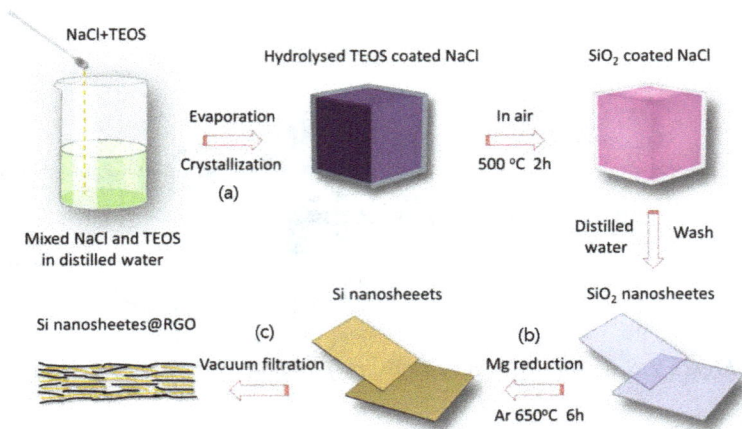

Figure 10. Schematic showing synthesis of silicon nanosheets using NaCl particles as template and the hydrolyzed TEOS as the silica precursor followed by magnesiothermic reduction. (Adopted with permission from ref [120]. ©2018 Elsevier)

It is hypothesized that large surface area of recrystallised NaCL contributes to the formation of large SiO_2 nanosheets via a self-assembly process on the surface of NaCl. The nanocomposites of the so produced silicon nanosheets with rGO exhibit cycling stability and improved rate performance.

1.2.4 3D-nanostructures

Several variants of 3D silicon nanostructures in the form of yolk-shell and core-shell morphologies have been explored. One of the most common approaches is the synthesis of carbon coated silicon in form of core-shell assemblies. Such an arrangement enables in overcoming the low conductivity and structural instability arising from large volume changes and high resistance of silicon [121–125]. The presence of carbon layer assists in decreasing the resistance of silicon particles significantly resulting in sustenance of reversible capacity of Li-Si alloys. However, the breaking of carbon shell in such an arrangement seriously hampers the reversible Li extraction which not only results in increased electrode resistance but also leads to fading of the reversible capacity. The movability of the core in a yolk-shell structure in a silicon/carbon system enables in providing space for more volume expansion and provides an effective solution for volume expansion during the lithiation process. In a recent development Si micro particles were encapsulated in conformally synthesized multi-layered graphene cages. The graphene cage allows the silicon micro particles to fracture and performs as a flexible buffer during cycling thereby retaining the electrical conductivity of the material. This enabled in achieving a high tap density along with high cycling stability [126]. It has also been observed in independent investigations that chemically inert graphene layer reduces the irreversible consumption of lithium ions and rapidly increases the Coulombic efficiency by forming a stable SEI film as compared to pyrolytic carbon [127]. This is suggestive of much higher flexibility and electrical conductivity of graphene as compared to that of pyrolytic carbon. This high flexibility of graphene sheets effectively balances the volume strain resulting during the galvanostatic cycling while high electrical conductivity enables in decreasing the internal resistance of the cell [127,128].

The schematic in Fig. 11 shows the synthesis of the spongy nanographene functionalized Silicon (Si@SG) composite. [129].

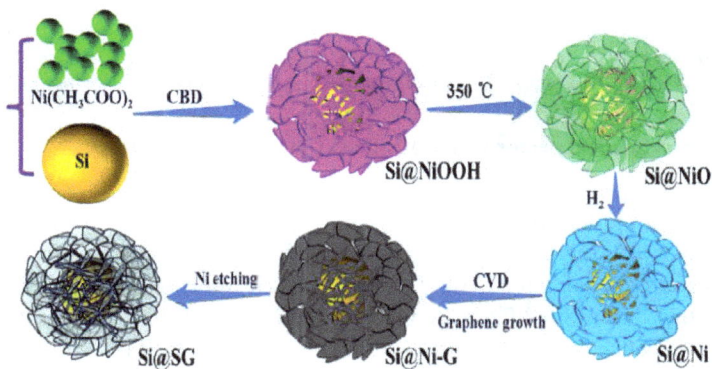

Figure 11. Schematic showing synthesis of Si@SG nanostructures. (Adopted with permission from ref [129]. ©2018 Springer)

The spongy nanographene shells were synthesized from CVD technique using nickel based templates. The presence of large number of graphene sheets stacked on silicon not only allows better electron conductivity but also allows fast migration of Li^+ and electrolyte ions near the silicon surface providing reduced mass barrier effects. The porous frameworks also allow to absorb the volume stress due to expansion of the silicon lattice during charging and discharging of the cell. These morphologies are different as compared to conventional mixture of rGO shells over silicon as shown in Fig. 12 (a) and (b). Thus this approach enables in mitigating the issues related to conductivity, volume expansion and mass transfer in silicon anodes.

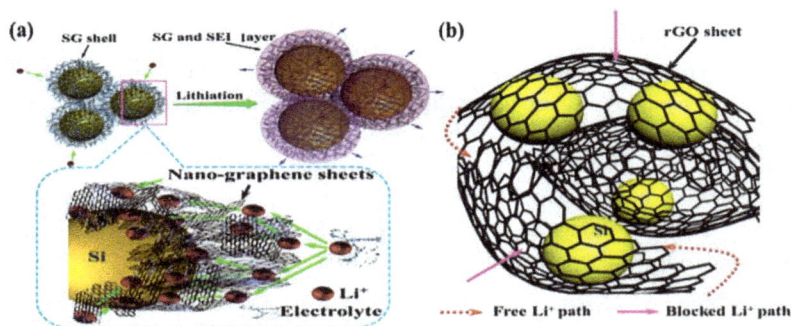

Figure 12. Schematic representation of (a) volume buffering (top) and mass transfer (bottom) mechanisms of the functional SG shell, (b) conventional mixture of rGO sheets and Si, (Adopted with permission from ref [129]. © 2018 Springer)

2.   Electrochemical performance of silicon based nanostructures

A comparative study on the performance of Si based nanostructures is evaluated using electrochemical testing of the anode material. These studies provide comprehensive understanding of charge transfer mechanisms and structural stability of nanostructures during the electrode cycling. Further various research works on synthesis of composite nanostructures of Si with different carbon materials, as well as the effects of metallic and non-metallic doping is also discussed. Surface modification of Si is one such strategy to strengthen the core of the Si nanomaterial to inhibit volume expansion and augment the mechanical stability to the material. In this technique, both ionically and electrically conductive layer is coated to prevent the electrolyte from persistent decomposition on the anode.

The galvanostatic voltage profiles of prepared Si with Prussian blue nano-architectures (PBAs). PBAs@Si and PBAs@Si-450 (sample treated at 450°C), as well as pure Si is shown in Fig. 13(a&b). The initial discharge capacities obtained from the galvanostatic profiles for pure Si, PBAs@Si and PBAs@Si-450 are approximately 3510mAh g^{-1}, 1899mAh g^{-1} and 2155mAh g^{-1}respectively. The initial Coulombic efficiencies for the sample treated at 450°C is higher compared to PBAs@Si-i.e; Coulombic efficiencies are in the order of, pure Si>PBAs@Si>PBAs@Si-450. Thus, the reduction in initial specific capacities and Coulombic efficiencies are mainly attributed to the reduction in silicon content. The annealing process helps in effective nitrogen doping of the samples. It is also hypothesized that the reduction in Coulombic efficiency of the nitrogen doped sample (PBAs@Si-450) is comparatively less which could be ascribed due to electrolytic degradation.

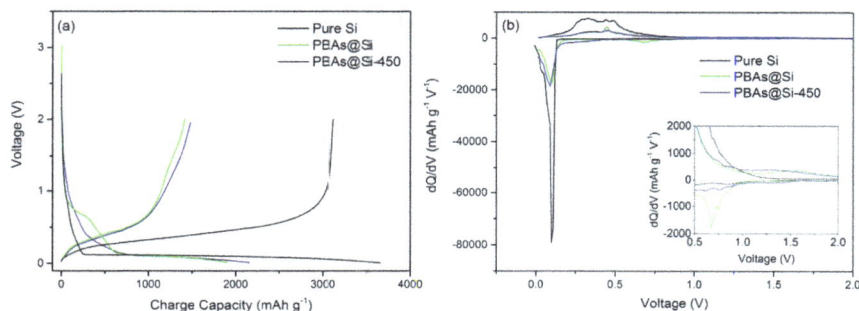

Figure 13 (a) Galvanostatic voltage profiles and (b) dQ/dV curves of pure Si, PBAs@Si, and PBAs@Si-450 for the first cycle at 0.07 A g⁻¹. (Adopted with permission from ref [93]. © 2019 ACS)

The irreversible loss in capacity is also observed as SEI peak in the dQ/dV plots shown in Fig. 14 (a&b). The trivial irreversible oxidation reactions occurring are realized from SEI layer formation and also from the formation of crystalline Si during the de-lithiation process of amorphous Li_xSi alloys [130]. Besides, the reduction peaks are accounted for the reversible lithiation of silicon and carbon /transition metals in PBAs derived shell [131]. The broad peak (0.6~0.8 V appears for PBAs@Si) during the first cycle suggests the formation of a SEI film layer on surface of the active material [132]. The other two subsequent cycles for PBAs@Si and PBAs@Si-450, in the dQ/dV curves are indicative of better cyclic reversibility. Further no significant peak shift is observed in the curves over 200 cycles which shows the enhanced cycling stability. The presence of mesoporous nature of the PBA-derived metal atoms and the SEI layer formation during nitrogen-doping over the carbon can be accounted for the stable behaviour. It is also unambiguous that the loss mechanism for PBAs@Si and PBAs@Si-450 is mechanical loss of the active material. The PBAs@Si-450 shows an outstanding rate capacity and structural stability (Fig.14c), a reversible charge capacity of 1450 mAh g⁻¹ is attained at a current density of 0.14 A g⁻¹ and even at a current density of 0.42 A g⁻¹ a specific charge capacity of 1136 mAh g^{-1} is recovered. Further the electrode kinetics is evaluated using the electrochemical impedance spectroscopy (EIS), as shown from Fig. 14d. It is apparent that the electrolytic and charge transfer resistances (R_{ct}) for pure Si electrode is higher compared to the other samples. Even though the R_{ct} of PBAs@Si-450 is greater than PBAs@Si during first cycle, subsequently after 200 cycles the charge transfer resistance is reduced. This mainly occurs due to the electrode cracking process or the presence of active particles which enhance the number of active sites for the electrochemical reactions.

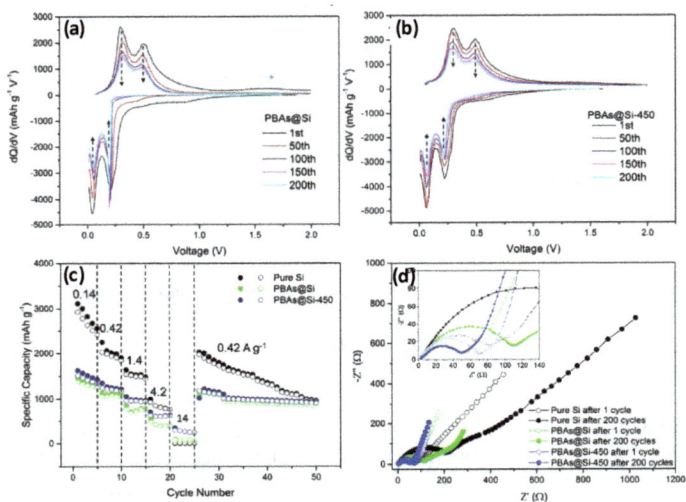

Figure 14. Electrochemical performances of pure Si, PBAs@Si and PBAs@Si-450. (a, b) dQ/dV curves of PBAs@Si and PBAs@Si-450 at 1st, 50th, 100th, 150th, 200th cycle. (c) Rate performance of pure Si, PBAs@Si and PBAs@Si-450 at various current densities ranging from 0.14, to 14 A g^{-1} and finally back to 0.42 A g^{-1}. (d) EIS plots of pure Si, PBAs@Si and PBAs@Si-450 after 1 cycle and after 200 cycles. (Adopted with permission from ref [93].© 2019 ACS)

Fig. 15 represents SEM micrographs of silver assisted chemically etched Si nanowires for an etching time of 10 min with a depth of 6 µm. The Si NWs and Ag particles (at the root regions) can be seen in the Fig. 15 inset. It is also seen that the etch front is not even, but it is likely that there is large distribution of shapes and different alignments of the Ag nano particles which is due to the fact that the contact area of the Ag particles defines the etching profile [133]. It is also imperative that this effect may also arise when an Ag film is used instead of Ag particles. The space between the silver nanoparticles also results in nanowires, whose thickness is in the range of 90 nm - 300 nm. Moreover, Si micropillar scan also be seen with some porous sites at their top. This has happened because of the lateral etching of residual silver nanoparticles with sizes below 50 nm.

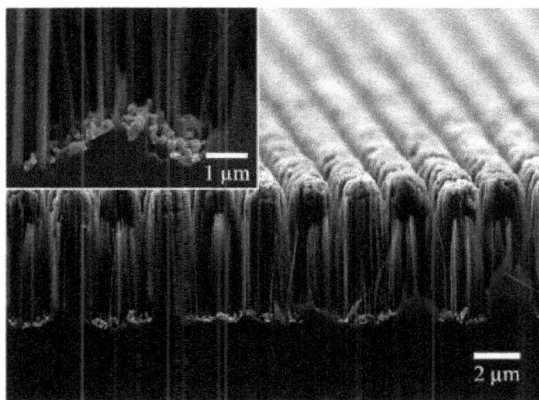

Figure 15. SEM micrograph of a sample after the etching process. Inset shows zoomed image of the formed Si NWs, with Ag particles at the bottom. (Adopted with permission from [135]. © 2019 Springer)

The walls of the Si micropillars present roughness given by the shape of the Ag particles. By means of a longer etching duration, micropillars with smoother walls can be yielded but the top most regions would be dissolved.

Similarly, the surface morphologies (Fig. 16 (a&b)) and the internal structure of 1D porous Si nanowires (PSiNWs)(Fig. 16 (c-f)) is shown which gives the comprehensive relation with the fabricated nanowires and their electrochemical properties. The cross-sectional surface in Fig.16a suggests that the 1D-PSiNW arrays are uniform over the entire surface. The magnified image (Fig. 16b) suggests that the width of perpendicularly arranged arrays of SiNWs in the range of 60.0-500.0 nm has mesoporous openings which are evenly spread over the length of each nanowire. Even after treatment with hydrofluoric acid the length of the as-prepared 1D-PSiNWs are still ~7.0 μm. Transmission electron microscopy (TEM) images (Fig.16(c-f)) shows that the NWs are highly porous at the surface, with both pore diameter and wall thickness around 7 nm. The obtained selected area electron diffraction (SAED) patterns (Fig.16c) and lattice fringes presented (Fig.16f) reveal that the porous NWs has polycrystalline structure and monocrystalline properties as well.

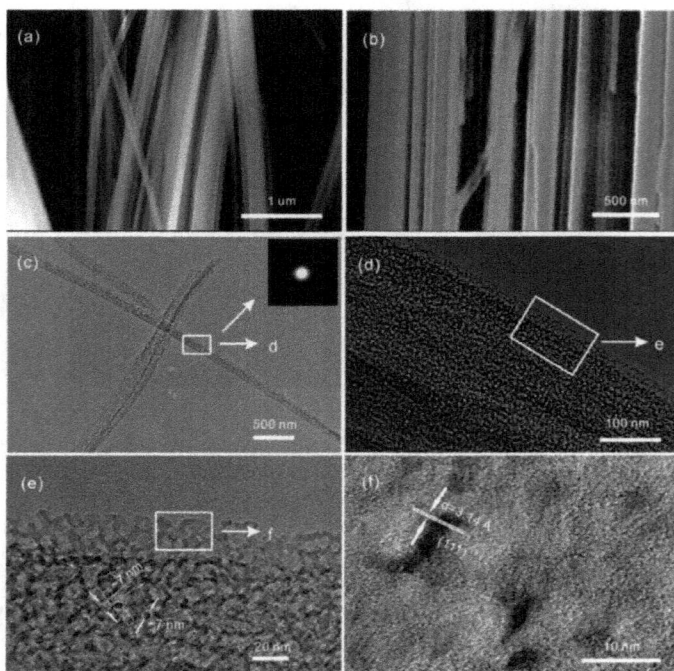

Figure 16. SEM (a,b), TEM(c,d), and HRTEM (e,f) images of 1D-PSiNW arrays etched. (Adopted with permission from ref [136]. ©2018 MDPI)

The electrochemical behaviour of 1D-PSiNWs were analysed using charge-discharge profiles as shown in Fig.17a. The initial capacities obtained during the discharge and charge for the anodes are 4487.2 mAh·g^{-1} and 2534.5 mAh·g^{-1} respectively, with 56.5% initial Coulombic efficiency. The irreversible loss of capacity was resulted due to SEI layer formation on the surfaces of the nanowires. Further to investigate stability of nanowires, cycle life performance is performed (Fig.17b) at various current density values. It turns out that nanowires exhibited a high capacity of 2061 mAh·g^{-1} over a thousand cycles. The Coulombic efficiency was much greater (99.7%) with a rate capability over a voltage window of 0.01V -2.0 V (Fig.17c). The specific capacity obtained during discharge is about 2706.5 mAh·g^{-1} at 0.4A·g^{-1} while retaining the capacity of 1667.0 mAh·g^{-1} at 4 A·g^{-1}. After cycling for 5000 cycles the anode still holds a higher specific capacity (585.7 mAh·g^{-1}) when compared to graphite (Capacity ~372.0 mAh·g^{-1}) (Fig. 17d).

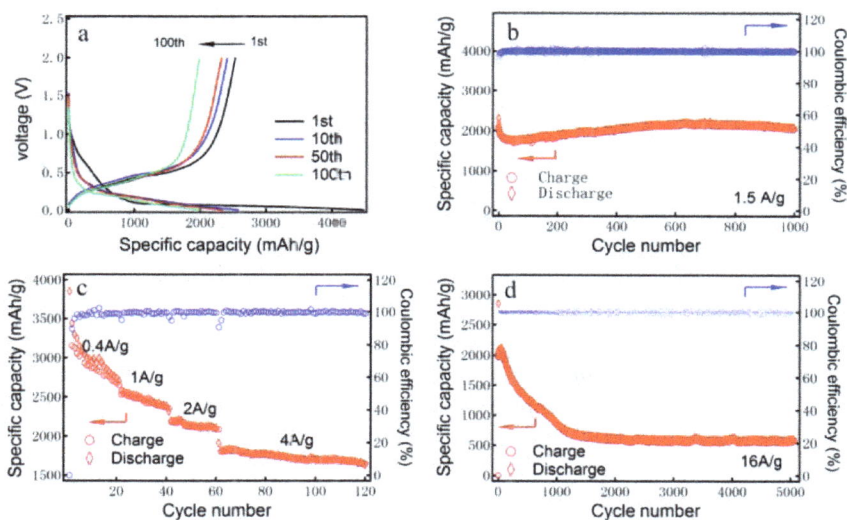

Figure 17. Results of electrochemical performance for 1D-PSiNWs anodes. (a) Galvanostatic charge/discharge profiles between 0.01V and 2.0 V vs. Li/Li$^+$ for the first, 10th, 50th, 100th cycles at a current densities of 1.0 Ag^{-1}. (b) cycling performance of the as-prepared 1DPSiNWs anodes at a current densties of 1.5 Ag^{-1} (c) rate performance of as prepared 1DPSiNWs anodes at a current densties of 0.4/1/2/4 Ag^{-1} (d) electrochemical performance of the as-prepared of 1DPSiNWs anodes with 5000 cycles at current densities of 16 Ag^{-1}.(Adopted with permission from ref [136].© 2018 MDPI)

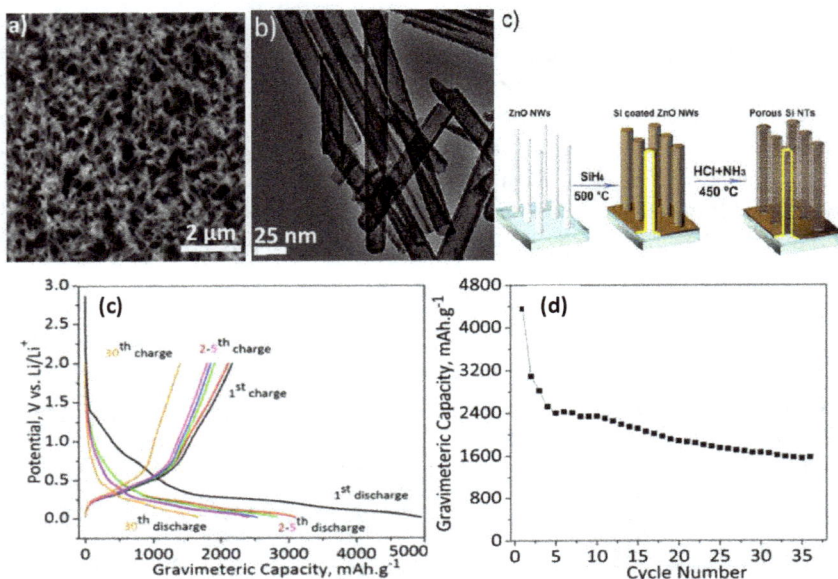

Figure 18. (a) SEM, (b) TEM image of Si NTAs and (c) Porous Si NTAs fabricated by three template directed method. (d) Galvanostatic charge-discharge profile performed and (e) cycle stability of Si NTAs performed at C/20. (Reprinted with permission from ref [134]. ©2017The Electrochemical Society)

Some other alternative routes to synthesize ultrathin crystalline silicon nanotube arrays (Si NTAs) were adopted using a three step template directed method [134]. The schematic process for synthesis of Si NTAs is also shown in Fig. 18c. The morphological studies (SEM and TEM shown in Fig. 18 (a&b respectively)) gives a clear understanding of the vertical growth of nanotube arrays. The Galvanostatic charge-discharge profile and cycle stability of the Si NTAs performed at C/20 in a potential window of 0.01 - 1.75 V vs Li/Li$^+$, respectively (Fig. 18d and 18e). The discharge profiles obtained for the SiNTAs show four small plateaus at 1.2 V, 0.7 V, 0.5 V and 0.4 V. These plateaus correspond to Li$^+$ reaction with electrolyte decomposition species to form SEI interface, a reaction which involves iron oxide present in stainless steel (SS) and alloying reaction of silicon. While the regions at 0.38 V and 0.5 V during initial charging process corresponds to the de-alloying mechanism of silicon. The Si NTAs/Li cell shows a large discharge capacity of 3095 mAh.g^{-1} which is due to the 3D porous structure that can accommodate large volume changes during cycling. The irreversible reaction of Li^+ with FeO and ZnO sacrificial template (not completely dissolved) used for synthesis process of the NTAs leads to a large fade in discharge capacity. During the cyclic stability test a gravimetric capacity of 1670 mAh.g^{-1} was obtained after 30 cycles.

Graphene coated templates are also widely used nowadays in the field of energy storage as graphene provides high surface enhancement. The Si coated graphene structures are also considered to be potential candidates with excellent stability during cycling and also a right choice to solve the degradation issues of silicon [137, 138]. The structure of silica coated graphene (Si@SG) shown in Fig.19(a). The micrograph reveals that each silicon particle is individually wrapped in the sponge like nanographene (SG) matrix. It resembles a flower-like structure of the Ni nanosheet template retaining several layers of nanographene sheets as shown in Fig.11. The Fig.19(b) shows the diffraction fringes whose interlayer spacing is calculated to be approximately 3.5 Å. Further the electrochemical studies on graphene composite matrix were investigated by fabricating a coin cell. It delivers a high reversible discharge capacity of 2330 mAh·g^{-1} (2532 mAh·g^{-1} when considering silicon alone) a: 0.250 A·g^{-1} and retains a high capacity of 1385 mAh·g^{-1} (1505 mAh·g^{-1} with silicon alone) after 510 cycles at 0.500 A·g^{-1} as shown in Fig.19(c). It suggests superior cycling performance with the hierarchical graphene matrix (Fig. 19(d)). The capacity obtained during the discharge is about 1470 mAh·g^{-1} is obtained after twenty cycles. The fade in capacity is only 5% even after 510 cycles. Moreover, the Coulombic efficiency (CE) reported for the Si@SG composite is 83.4% for the first cycle, denoting an excellent initial CE, which is rarely reported elsewhere [139].

As far as material conductivity is considered reduced graphene oxide can be used to increase the conductivity of anodes [140]. Compared with most conventional rGO-modified silicon composite electrodes (obtained by physical mixing or chemical modification), the as-synthesized Si@SG composite shows a much higher cycling stability even if at a higher charge-discharge current density, which may represent a strong indication that the new porous nanographene shell possesses both an excellent volume-buffering ability and a lower mass (Li^+) transfer resistance. However, among the reported Si/graphene composites, only silicon micrometer particles wrapped by both carbon and rGO achieves almost zero capacity loss after 160 cycles [141]. Similarly, silicon nano particles encapsulated with nitrogen doped rGO graphene and carbon nanotubes exhibits high specific capacity and good stability [142]. This suggests that a further contribution to the improved capacity, in addition to the effects of the higher conductivity and the more stable SEI layer, may also come from the carbon component.

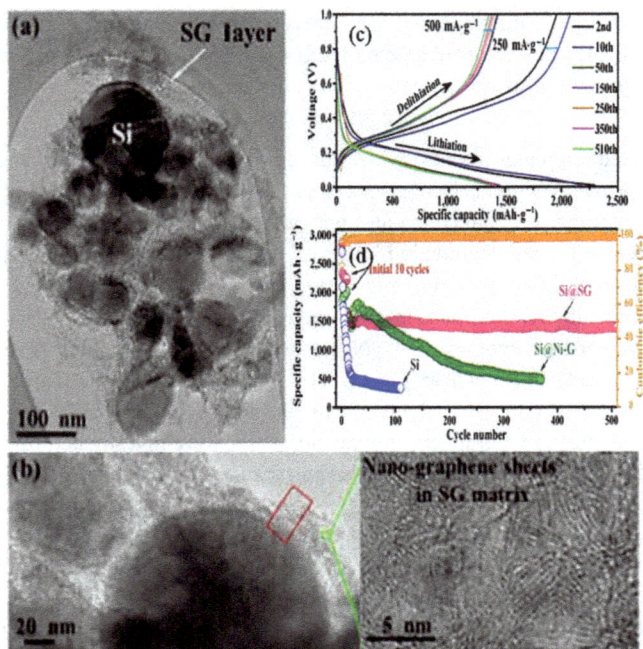

Figure 19 (a) TEM image of monolithic Si@SG composite, (b) High-magnification TEM image of individual silicon unit wrapped with sponge-like nanographene (left), and High-resolution TEM images of the edge (right), (c) Charge–discharge curves corresponding to the initial, 2nd, and 10th cycles at a low current density of 250 mA·g^{-1}, and to the 50th, 150th, 250th, 350th, and 510th cycles at a high current density of 500 mA·g^{-1} and (d) Cycling performance (lithiation capacity) of Si, Si@Ni-G, and Si@SG samples at a current density of 500 mA·g^{-1}, and CE of Si@SG composite measured during 510 cycles. All electrodes were initially cycled at 250 mA·g^{-1} (for the first 10 cycles) and then at 500 mA·g^{-1} in subsequent cycles. (Adopted with permission from ref [129]. © 2018 Springer)

In comparison to various nanostructured composites of Si based electrodes reviewed in this work so far, it can be asserted that the materials with the highest capacities are not necessarily the best implementation in a real lithium ion battery. Further there is no exact correlation between performance and specific dimensionality; however, it seems as if the use of carbon, especially in the form of CNTs and graphene as a conductive and supportive matrix, tends to improve the cycling stability and Coulombic efficiency. Also, composites and hierarchical structures of Si based anodes seem to exhibit a very stable performance compared to simple structured approaches. While structures of the latter category are often easier to synthesize, it still might be more fruitful to further explore the possibilities of Si/C based hierarchical composite structures. In fact, most recent works actually tend to go down that road. It should be noted that there won't be one anode that suits all needs and purposes, but rather different materials for either high capacities, ultra-long lifetime or high energy output. This is also reflected in the summary of the materials reviewed in this work (Table 1), where each material shows specific weaknesses and strengths. Finally, despite all the challenges and problems that still need to be overcome, silicon-based anode materials are one of the most promising anodes for the next generation Li-ion batteries and would enormously boost their performance compared to the current cells.

Table 1. Comparison showing various nanostructures of Si and its composites in all the three dimensions and their performance as anode material for Li battery

Material	Dimension	Initial charge capacity (mAhg⁻¹)	Initial discharge capacity (mAhg⁻¹)	Initial CE	Reversible capacity (mAhg⁻¹)@Current density @(mAg⁻¹)	Cycle number	Rate capability capacity @(mAhg⁻¹) @Current density @(mAg⁻¹)	CE	Ref
Pure Si NPs	0D	4189	3549	84.7	1180@3000	500	1180@3000	>98	144
Non filling carbon coated porous Si micrometer sized particles	0D	1798	NA	78	1490@1050	1000	NA	NA	141
Porous Si Nws with C coating	1D	NA	3600	NA	1500@400	50	NA	97	145
C@Si@C coreshell Si nanotube	1D	1565	1780	88	700@3000	300	600@4000	99.95	146
Sandwiched structured C/Si/TiO₂ nanofiber composite	1D	946	1771	53.4	793@100	160	176@3000	>98	147
Mesoporous Si nanosheets wrapped in carbon	2D	1658	3474	47.7	1072@4000	500	1297@4000	~100	148
Si nanoparticles wrapped into 2D carbon nanosheets	2D	NA	1723	70.4	881@420	500	401@3100	>99	149
Porous Si with carbon wrapping	3D	NA	4224	66.8	1639@1000	200	1074@400■	>98	150
Structure preserved 3D porous Si/rGO	3D	1563	NA	60.1	1210@500	100	955@2000	>99	151

Conclusion

Silicon is extensively used as an anode material for the lithium-ion battery technologies. The high theoretical capacity makes it an interesting material to be widely used in battery applications. From the past few decades, there is tremendous research on nanostructured materials to address the serious problems associated with the Si anode volume expansion and to obtain a stable capacity. However, its commercial application at industrial level is still limited due to several issues related to its nanostructures.

References

[1] B. Dunn, H. Kamath, J.-M. Tarascon, Electrical energy storage for the grid: A battery of choices, Science. 334 (2011) 928-935. https://doi.org/10.1126/science.1212741

[2] N. Kittner, F. Lill, D.M. Kammen, Energy storage deployment and innovation for the clean energy transition, Nat. Energy. 2 (2017) 17125. https://doi.org/10.1038/nenergy.2017.125

[3] A. Yoshino, The birth of the lithium-ion battery, Angew. Chemie Int. Ed. 51 (2012) 5798-5800. https://doi.org/10.1002/anie.201105006

[4] G. Assat, J.M. Tarascon, Fundamental understanding and practical challenges of anionic redox activity in Li-ion batteries, Nat. Energy. 3 (2018) 373-386. https://doi.org/10.1038/s41560-018-0097-0

[5] H.L. Tuller, Solar to fuels conversion technologies: a perspective, Mater. Renew. Sustain. Energy. 6 (2017) 3. https://doi.org/10.1007/s40243-017-0088-2

[6] M. Padmini, S.K. Kiran, N. Lakshminarasimhan, M. Sathish, P. Elumalai, High-performance solid-state hybrid energy-storage device consisting of reduced graphene-oxide anchored with Ni Mn-layered double hydroxide, Electrochim. Acta. 236 (2017) 359-370. https://doi.org/10.1016/j.electacta.2017.03.170

[7] P.A. Owusu, S. Asumadu-Sarkodie, A review of renewable energy sources, sustainability issues and climate change mitigation, Cogent Eng. 3 (2016) 1167990. https://doi.org/10.1080/23311916.2016.1167990

[8] K. K. Sarigamala, S. Shukla, A.Struck, S. Saxena, Rationally engineered 3D-dendritic cell-like morphologies of LDH nanostructures using graphene based core-shell structures, Microsyst. Nanoeng 5 (2019) 65. https://doi.org/10.1038/s41378-019-0114-x

[9] M. Armand, J.-M. Tarascon, Building better batteries, Nature. 451 (2008) 652-657. https://doi.org/10.1038/451652a

[10] D.W.H. Lambert, J.E. Manders, R.F. Nelson, K. Peters, D.A.J. Rand, M. Stevenson, Strategies for enhancing lead-acid battery production and performance, J. Power Sources. 88 (2000) 130-147. https://doi.org/10.1016/S0378-7753(99)00521-2

[11] U. Köhler, C. Antonius, P. Bäuerlein, Advances in alkaline batteries, J. Power Sources. 127 (2004) 45-52. https://doi.org/10.1016/j.jpowsour.2003.09.006

[12] M. Li, J. Lu, Z. Chen, K. Amine, 30 Years of lithium-ion batteries, Adv. Mater. 30 (2018) 1800561. https://doi.org/10.1002/adma.201800561

[13] X. Yu, A. Manthiram, Electrode-electrolyte interfaces in lithium-based batteries, Energy Environ. Sci. 11 (2018) 527-543. https://doi.org/10.1039/C7EE02555F

[14] Y. Sun, N. Liu, Y. Cui, Promises and challenges of nanomaterials for lithium-based rechargeable batteries, Nat. Energy. 1 (2016) 16071. https://doi.org/10.1038/nenergy.2016.71

[15] M.S. Whittingham, Electrical Energy Storage and Intercalation Chemistry, Science 192 (1976) 1126-1127. https://doi.org/10.1126/science.192.4244.1126

[16] Bottled lightning: superbatteries, electric cars, and the new lithium economy, Choice Rev. Online. 49 (2011) 49-1488. https://doi.org/10.5860/CHOICE.49-1488

[17] B. Lung-Hao Hu, F.-Y. Wu, C.-T. Lin, A.N. Khlobystov, L.-J. Li, Graphene-modified LiFePO4 cathode for lithium ion battery beyond theoretical capacity, Nat. Commun. 4 (2013) 1687. https://doi.org/10.1038/ncomms2705

[18] B. Scrosati, J. Garche, Lithium batteries: Status, prospects and future, J. Power Sources. 195 (2010) 2419-2430 https://doi.org/10.1016/j.jpowsour.2009.11.048

[19] P.L. Taberna, S. Mitra, P. Poizot, P. Simon, J.M. Tarascon, High rate capabilities Fe3O4-based Cu nano-architectured electrodes for lithium-ion battery applications, Nat. Mater. 5 (2006) 567-573. https://doi.org/10.1038/nmat1672

[20] M. Gauthier, T.J. Carney, A. Grimaud, L. Giordano, N. Pour, H.-H. Chang, D.P. Fenning, S.F. Lux, O. Paschos, C. Bauer, F. Maglia, S. Lupart, P. Lamp, Y. Shao-Horn, Electrode-electrolyte interface in Li-ion batteries: Current understanding and new insights, J. Phys. Chem. Lett. 6 (2015) 4653-4672. https://doi.org/10.1021/acs.jpclett.5b01727

[21] X.-L. Gou, J. Chen, P.-W. Shen, Synthesis, characterization and application of SnSx (x=1,2) nanoparticles, Mater. Chem. Phys. 93 (2005) 557-563. https://doi.org/10.1016/j.matchemphys.2005.04.008

[22] P. Suresh, A.K. Shukla, N. Munichandraiah, Electrochemical properties of LiMn1−xMxO2 (M=Ni, Al, Mg) as cathode materials in lithium-ion cells, J. Electrochem. Soc. 152 (2005) A2273. https://doi.org/10.1149/1.2073067

[23]Z. Lu, D.D. MacNeil, J.R. Dahn, Layered LiNixCo1−2xMnxO2 cathode materials for lithium-ion batteries, Electrochem. Solid-State Lett. 4 (2001) A200. https://doi.org/10.1149/1.1413182

[24] M.R. Mancini, L. Petrucci, F. Ronci, P.P. Prosini, S. Passerini, Long cycle life Li-Mn-O defective spinel electrodes, J. Power Sources. 76 (1998) 91. https://doi.org/10.1016/S0378-7753(98)00144-X

[25] S.-H. Wu, K.-M. Hsiao, W.-R. Liu, The preparation and characterization of olivine LiFePO4 by a solution method, J. Power Sources. 146 (2005) 550-554. https://doi.org/10.1016/j.jpowsour.2005.03.128

[26] S. S. Anish, S. Saxena, P. Shrivastava, S. Shukla,Looking beyond single electron extraction in cathode materials for lithium ion batteries, J. Power Sources 279 (2015) 563-566. https://doi.org/10.1016/j.jpowsour.2015.01.061

[27] Z.-S. Wu, G. Zhou, L.-C. Yin, W. Ren, F. Li, H.-M. Cheng, Graphene/metal oxide composite electrode materials for energy storage, Nano Energy. 1 (2012) 107-131. https://doi.org/10.1016/j.nanoen.2011.11.001

[28] D. Zhang, Graphene enhanced LiFeBO3/C composites as cathodes for Li- ion batteries, Int. J. Electrochem. Sci. (2018) 1744-1753. https://doi.org/10.20964/2018.02.52

[29] H. Bin Wu, J.S. Chen, H.H. Hng, X. Wen (David) Lou, Nanostructured metal oxide-based materials as advanced anodes for lithium-ion batteries, Nanoscale. 4 (2012) 2526. https://doi.org/10.1039/c2nr11966h

[30] Z.-L. Xu, X. Liu, Y. Luo, L. Zhou, J.-K. Kim, Nanosilicon anodes for high performance rechargeable batteries, Prog. Mater. Sci. 90 (2017) 1-44. https://doi.org/10.1016/j.pmatsci.2017.07.003

[31] N. Nitta, F. Wu, J.T. Lee, G. Yushin, Li-ion battery materials: present and future, Mater. Today. 18 (2015) 252-264. https://doi.org/10.1016/j.mattod.2014.10.040

[32]B. Fuchsbichler, C. Stangl, H. Kren, F. Uhlig, S. Koller, High capacity graphite-silicon composite anode material for lithium-ion batteries, J. Power Sources. 196 (2011) 2889-2892. https://doi.org/10.1016/j.jpowsour.2010.10.081

[33] B. Sun, Z. Chen, H.S. Kim, H. Ahn, G. Wang, MnO/C core-shell nanorods as high capacity anode materials for lithium-ion batteries, J. Power Sources. 196 (2011) 3346-3349. https://doi.org/10.1016/j.jpowsour.2010.11.090

[34] S. Goriparti, E. Miele, F. De Angelis, E. Di Fabrizio, R. Proietti Zaccaria, C. Capiglia, Review on recent progress of nanostructured anode materials for Li-ion batteries, J. Power Sources. 257 (2014) 421-443. https://doi.org/10.1016/j.jpowsour.2013.11.103

[35] S. Yin, Q. Ji, X. Zuo, S. Xie, K. Fang, Y. Xia, J. Li, B. Qiu, M. Wang, J. Ban, X. Wang, Y. Zhang, Y. Xiao, L. Zheng, S. Liang, Z. Liu, C. Wang, Y.-J. Cheng, Silicon lithium-ion battery anode with enhanced performance: Multiple effects of silver nanoparticles, J. Mater. Sci. Technol. 34 (2018) 1902-1911. https://doi.org/10.1016/j.jmst.2013.02.004

[36] M.J. Loveridge, M.J. Lain, I.D. Johnson, A. Roberts, S.D. Beattie, R. Dashwood, J.A. Darr, R. Bhagat, Towards high capacity Li-ion batteries based on silicon-graphene composite anodes and sub-micron V-doped LiFePO4 cathodes, Sci. Rep. 6 (2016) 37787. https://doi.org/10.1038/srep37787

[37] C. Liu, Z.G. Neale, G. Cao, Understanding electrochemical potentials of cathode materials in rechargeable batteries, Mater. Today. 19 (2016) 109-123. https://doi.org/10.1016/j.mattod.2015.10.009

[38] J. Wen, Y. Yu, C. Chen, A Review on Lithium-ion batteries safety issues: Existing problems and possible solutions, Mater. Express. 2 (2012) 197-212. https://doi.org/10.1166/mex.2012.1075

[39] J.Y. Li, Q. Xu, G. Li, Y.X. Yin, L.J. Wan, Y.G. Guo, Research progress regarding Si-based anode materials towards practical application in high energy density Li-ion batteries, Mater. Chem. Front. 1 (2017) 1691-1708. https://doi.org/10.1039/C6QM00302H

[40] N. Nitta, G. Yushin, High-capacity anode materials for lithium-ion batteries: Choice of elements and structures for active particles, Part. Part. Syst. Charact. 31 (2014) 317-336. https://doi.org/10.1002/ppsc.201300231

[41] A. Bordes, E. De Vito, C. Haon, A. Boulineau, A. Montani, P. Marcus, Multiscale investigation of silicon anode Li insertion mechanisms by time-of-flight secondary ion mass spectrometer imaging performed on an in situ focused ion beam cross section, Chem. Mater. 28 (2016) 1566-1573. https://doi.org/10.1021/acs.chemmater.6b00155

[42] M.T. McDowell, S.W. Lee, J.T. Harris, B.A. Korgel, C. Wang, W.D. Nix, Y. Cui, In situ TEM of two-phase lithiation of amorphous silicon nanospheres, Nano Lett. 13 (2013) 758-764. https://doi.org/10.1021/nl3044508

[43] H. Wu, Y. Cui, Designing nanostructured Si anodes for high energy lithium ion batteries, Nano Today. 7 (2012) 414-429. https://doi.org/10.1016/j.nantod.2012.08.004

[44] S.W. Lee, M.T. McDowell, L.A. Berla, W.D. Nix, Y. Cui, Fracture of crystalline silicon nanopillars during electrochemical lithium insertion, Proc. Natl. Acad. Sci. 109 (2012) 4080-4085. https://doi.org/10.1073/pnas.1201088109

[45] X.H. Liu, L. Zhong, S. Huang, S.X. Mao, T. Zhu, J.Y. Huang, Size-dependent fracture of silicon nanoparticles during lithiation, ACS Nano. 6 (2012) 1522-1531. https://doi.org/10.1021/nn204476h

[46]P. Verma, P. Maire, P. Novák, A review of the features and analyses of the solid electrolyte interphase in Li-ion batteries, Electrochim. Acta. 55 (2010) 6332-6341. https://doi.org/10.1016/j.electacta.2010.05.072

[47] A.L. Michan, G. Divitini, A.J. Pell, M. Leskes, C. Ducati, C.P. Grey, Solid electrolyte interphase growth and capacity loss in silicon electrodes, J. Am. Chem. Soc. 138 (2016) 7918-7931. https://doi.org/10.1021/jacs.6b02882

[48] J.P. Yen, C.C. Chang, Y.R. Lin, S.T. Shen, J.L. Hong, Sputtered copper coating on silicon/graphite composite anode for lithium ion batteries, J. Alloys Compd. 598 (2014) 184-190. https://doi.org/10.1016/j.jallcom.2014.01.230

[49] M. Ramesh, H.S. Nagaraja, Effect of current density on morphological, structural and optical properties of porous silicon, Mater. Today Chem. 3 (2017) 10-14. https://doi.org/10.1016/j.mtchem.2016.12.002

[50] M. Thakur, M. Isaacson, S.L. Sinsabaugh, M.S. Wong, S.L. Biswal, Gold-coated porous silicon films as anodes for lithium ion batteries, J. Power Sources. 205 (2012) 426-432. https://doi.org/10.1016/j.jpowsour.2012.01.058

[51] E.M. Lotfabad, P. Kalisvaart, A. Kohandehghan, K. Cui, M. Kupsta, B. Farbod, D. Mitlin, Si nanotubes ALD coated with TiO2, TiN or Al 2 O 3 as high performance lithium ion battery anodes, J. Mater. Chem. A. 2 (2014) 2504-2516. https://doi.org/10.1039/C3TA14302C

[52] M. Ramesh, H.S. Nagaraja, The effect of etching time on structural properties of Porous silicon at the room temperature, Mater. Today Proc. 3 (2016) 2085-2090. https://doi.org/10.1016/j.matpr.2016.04.112

[53]Y. Yao, N. Liu, M.T. McDowell, M. Pasta, Y. Cui, Improving the cycling stability of silicon nanowire anodes with conducting polymer coatings, Energy Environ. Sci. 5 (2012) 7927. https://doi.org/10.1039/c2ee21437g

[54]D. Tang, R. Yi, M.L. Gordin, M. Melnyk, F. Dai, S. Chen, J. Song, D. Wang, Titanium nitride coating to enhance the performance of silicon nanoparticles as a lithium-ion battery anode, J. Mater. Chem. A. 2 (2014) 10375-10378. https://doi.org/10.1039/C4TA01343C

[55] N. Dimov, S. Kugino, M. Yoshio, Carbon-coated silicon as anode material for lithium ion batteries: Advantages and limitations, Electrochim. Acta. 48 (2003) 1579-1587. https://doi.org/10.1016/S0013-4686(03)00030-6

[56] M. Yoshio, H. Wang, K. Fukuda, T. Umeno, N. Dimov, Z. Ogumi, Carbon-coated Si as a lithium-ion battery anode material, J. Electrochem. Soc. 149 (2002) A1598. https://doi.org/10.1149/1.1518988

[57] I. Hasa, J. Hassoun, S. Passerini, Nanostructured Na-ion and Li-ion anodes for battery application: A comparative overview, Nano Res. 10 (2017) 3942-3969. https://doi.org/10.1007/s12274-017-1513-7

[58] X. Zuo, J. Zhu, P. Müller-Buschbaum, Y.-J. Cheng, Silicon based lithium-ion battery anodes: A chronicle perspective review, Nano Energy. 31 (2017) 113-143. https://doi.org/10.1016/j.nanoen.2016.11.013

[59] X.H. Liu, H. Zheng, L. Zhong, S. Huang, K. Karki, L.Q. Zhang, Y. Liu, A. Kushima, W.T. Liang, J.W. Wang, J.-H. Cho, E. Epstein, S.A. Dayeh, S.T. Picraux, T. Zhu, J. Li, J.P. Sullivan, J. Cumings, C. Wang, S.X. Mao, Z.Z. Ye, S. Zhang, J.Y. Huang, Anisotropic swelling and fracture of silicon nanowires during lithiation, Nano Lett. 11 (2011) 3312-3318. https://doi.org/10.1021/nl201684d

[60] L. Goldstein, F. Glas, J.Y. Marzin, M.N. Charasse, G. Le Roux, Growth by molecular beam epitaxy and characterization of InAs/GaAs strainedlayer superlattices, Appl. Phys. Lett. 47 (1985) 1099-1101. https://doi.org/10.1063/1.96342

[61] J. Stangl, V. Holý, G. Bauer, Structural properties of self-organized semiconductor nanostructures, Rev. Mod. Phys. 76 (2004) 725-783. https://doi.org/10.1103/RevModPhys.76.725

[62] C.M. Hessel, E.J. Henderson, J.G.C. Veinot, Hydrogen Silsesquioxane: A molecular precursor for nanocrystalline Si−SiO2 composites and freestanding hydride-surface-terminated silicon nanoparticles, Chem. Mater. 18 (2006) 6139-6146. https://doi.org/10.1021/cm0602803

[63] J. Ryu, D. Hong, M. Shin, S. Park, Multiscale hyperporous silicon flake anodes for high initial Coulombic efficiency and cycle stability, ACS Nano. 10 (2016) 10589-10597. https://doi.org/10.1021/acsnano.6b06828

[64] H. Sohn, D.H. Kim, R. Yi, D. Tang, S.E. Lee, Y.S. Jung, D. Wang, Semimicro-size agglomerate structured silicon-carbon composite as an anode material for high performance lithium-ion batteries, J. Power Sources. 334 (2016) 128-136. https://doi.org/10.1016/j.jpowsour.2016.09.096

[65] L. Wei, Z. Hou, H. Wei, Porous sandwiched graphene/silicon anodes for lithium storage, Electrochim. Acta. 229 (2017) 445-451. https://doi.org/10.1016/j.electacta.2017.01.173

[66] L. Wei, Z. Hou, High performance polymer binders inspired by chemical finishing of textiles for silicon anodes in lithium ion batteries, J. Mater. Chem. A. 5 (2017) 22156-22162. https://doi.org/10.1039/C7TA05195F

[67] N.-W. Li, Y.-X. Yin, S. Xin, J.-Y. Li, Y.-G. Guo, Methods for the stabilization of nanostructured electrode materials for advanced rechargeable batteries, Small Methods. 1 (2017) 1700094. https://doi.org/10.1002/smtd.201700094

[68] N. Liu, H. Wu, M.T. McDowell, Y. Yao, C. Wang, Y. Cui, A yolk-shell design for stabilized and scalable Li-ion battery alloy anodes, Nano Lett. 12 (2012) 3315-3321. https://doi.org/10.1021/nl3014814

[69] Q. Xu, J.-Y. Li, J.-K. Sun, Y.-X. Yin, L.-J. Wan, Y.-G. Guo, Watermelon-inspired Si/C microspheres with hierarchical buffer structures for densely compacted lithium-ion battery anodes, Adv. Energy Mater. 7 (2017) 1601481. https://doi.org/10.1002/aenm.201601481

[70] J. Xie, L. Tong, L. Su, Y. Xu, L. Wang, Y. Wang, Core-shell yolk-shell Si@C@Void@C nanohybrids as advanced lithium ion battery anodes with good electronic conductivity and corrosion resistance, J. Power Sources. 342 (2017) 529-536. https://doi.org/10.1016/j.jpowsour.2016.12.094

[71] D. Hong, J. Ryu, S. Shin, S. Park, Cost-effective approach for structural evolution of Si-based multicomponent for Li-ion battery anodes, J. Mater. Chem. A. 5 (2017) 2095-2101. https://doi.org/10.1039/C6TA08889A

[72] X. Zuo, Y. Xia, Q. Ji, X. Gao, S. Yin, M. Wang, X. Wang, B. Qiu, A. Wei, Z. Sun, Z. Liu, J. Zhu, Y.-J. Cheng, Self-Templating Construction of 3D hierarchical macro-/mesoporous silicon from 0D silica nanoparticles, ACS Nano. 11 (2017) 889-899. https://doi.org/10.1021/acsnano.6b07450

[73] S.J. Lee, H.J. Kim, T.H. Hwang, S. Choi, S.H. Park, E. Deniz, D.S. Jung, J.W. Choi, Delicate structural control of Si-SiOx-C composite via high-speed spray pyrolysis for Li-ion battery anodes, Nano Lett. 17 (2017) 1870-1876. https://doi.org/10.1021/acs.nanolett.6b05191

[74] M. Sohn, H.-I. Park, H. Kim, Foamed silicon particles as a high capacity anode material for lithium-ion batteries, Chem. Commun. 53 (2017) 11897-11900. https://doi.org/10.1039/C7CC06171D

[75] R. Zhang, Y. Du, D. Li, D. Shen, J. Yang, Z. Guo, H.K. Liu, A.A. Elzatahry, D. Zhao, Highly reversible and large lithium storage in mesoporous Si/C nanocomposite anodes with silicon nanoparticles embedded in a carbon framework, Adv. Mater. 26 (2014) 6749-6755. https://doi.org/10.1002/adma.201402813

[76] S. Guo, X. Hu, Y. Hou, Z. Wen, Tunable Synthesis of yolk-shell porous silicon@carbon for optimizing Si/C-based anode of lithium-ion batteries, ACS Appl. Mater. Interfaces. 9 (2017) 42084-42092. https://doi.org/10.1021/acsami.7b13035

[77] W. Luo, Y. Wang, L. Wang, W. Jiang, S.-L. Chou, S.X. Dou, H.K. Liu, J. Yang, Silicon/mesoporous Carbon/crystalline TiO2 nanoparticles for highly stable lithium storage, ACS Nano. 10 (2016) 10524-10532. https://doi.org/10.1021/acsnano.6b06517

[78] Y. Jin, S. Li, A. Kushima, X. Zheng, Y. Sun, J. Xie, J. Sun, W. Xue, G. Zhou, J. Wu, F. Shi, R. Zhang, Z. Zhu, K. So, Y. Cui, J. Li, Self-healing SEI enables full-cell cycling of a silicon-majority anode with a coulombic efficiency exceeding 99.9%, Energy Environ. Sci. 10 (2017) 580-592. https://doi.org/10.1039/C6EE02685K

[79] L.Y. Yang, H.Z. Li, J. Liu, Z.Q. Sun, S.S. Tang, M. Lei, Dual yolk-shell structure of carbon and silica-coated silicon for high-performance lithium-ion batteries, Sci. Rep. 5 (2015) 10908. https://doi.org/10.1038/srep10908

[80] Q. Xu, J.-K. Sun, Y.-X. Yin, Y.-G. Guo, Facile synthesis of blocky SiOx/C with graphite-like structure for high-performance lithium-ion battery anodes, Adv. Funct. Mater. 28 (2018) 1705235. https://doi.org/10.1002/adfm.201705235

[81] T. Jesionowski, Preparation of spherical silica in emulsion systems using the co-precipitation technique, Mater. Chem. Phys. 113 (2009) 839-849. https://doi.org/10.1016/j.matchemphys.2008.08.067

[82] R. Yuvakkumar, V. Elango, V. Rajendran, N. Kannan, High-purity nano silica powder from rice husk using a simple chemical method, J. Exp. Nanosci. 9 (2014) 272-281. https://doi.org/10.1080/17458080.2012.656709

[83] P. Nie, L. Shen, H. Luo, B. Ding, G. Xu, J. Wang, X. Zhang, Prussian blue analogues: a new class of anode materials for lithium ion batteries, J. Mater. Chem. A. 2 (2014) 5852-5857. https://doi.org/10.1039/C4TA00062E

[84] C.D. Wessells, R.A. Huggins, Y. Cui, Copper hexacyanoferrate battery electrodes with long cycle life and high power, Nat. Commun. 2 (2011) 550-554. https://doi.org/10.1038/ncomms1563

[85] J.-H. Lee, G. Ali, D.H. Kim, K.Y. Chung, Metal-organic framework cathodes based on a vanadium hexacyanoferrate Prussian Blue analogue for high-performance aqueous rechargeable batteries, Adv. Energy Mater. 7 (2017) 1601491. https://doi.org/10.1002/aenm.201601491

[86] R. Chen, Y. Huang, M. Xie, Z. Wang, Y. Ye, L. Li, F. Wu, Chemical inhibition method to synthesize highly crystalline Prussian Blue analogs for sodium-ion battery cathodes, ACS Appl. Mater. Interfaces. 8 (2016) 31669-31676. https://doi.org/10.1021/acsami.6b10884

[87] L. Ma, T. Chen, G. Zhu, Y. Hu, H. Lu, R. Chen, J. Liang, Z. Tie, Z. Jin, J. Liu, Pitaya-like microspheres derived from Prussian Blue analogues as ultralong-life anodes for lithium storage, J. Mater. Chem. A. 4 (2016) 15041-15048. https://doi.org/10.1039/C6TA06692E

[88] W. Zhang, Y. Zhao, V. Malgras, Q. Ji, D. Jiang, R. Qi, K. Ariga, Y. Yamauchi, J. Liu, J. Sen Jiang, M. Hu, Synthesis of monocrystalline nanoframes of Prussian Blue analogues by controlled preferential etching, Angew. Chemie - Int. Ed. 55 (2016) 8228-8234. https://doi.org/10.1002/anie.201600661

[89] Y. Lu, L. Wang, J. Cheng, J.B. Goodenough, Prussian Blue: A new framework of electrode materials for sodium batteries, Chem. Commun. 48 (2012) 6544. https://doi.org/10.1039/c2cc31777j

[90] P. Xiong, G. Zeng, L. Zeng, M. Wei, Prussian Blue analogues Mn[Fe(CN)6]0.6667·nH2O cubes as an anode material for lithium-ion batteries, Dalt. Trans. 44 (2015) 16746. https://doi.org/10.1039/C5DT03030G

[91] F. Ma, Q. Li, T. Wang, H. Zhang, G. Wu, Energy storage materials derived from Prussian Blue analogues, Sci. Bull. 62 (2017) 358-368. https://doi.org/10.1016/j.scib.2017.01.030

[92] L. Guo, R. Mo, W. Shi, Y. Huang, Z.Y. Leong, M. Ding, F. Chen, H.Y. Yang, A Prussian Blue anode for high performance electrochemical deionization promoted by the faradaic mechanism, Nanoscale. 9 (2017) 13305-13312. https://doi.org/10.1039/C7NR03579A

[93] F. Wu, H. Wang, J. Shi, Z. Yan, S. Song, B. Peng, X. Zhang, Y. Xiang, Surface modification of silicon nanoparticles by an "ink" layer for advanced lithium ion batteries, ACS Appl. Mater. Interfaces. 10 (2018) 19639. https://doi.org/10.1021/acsami.8b03000

[94] R. Martha, N. H.S., Effect of current density and electrochemical cycling on physical properties of silicon nanowires as anode for lithium ion battery, Mater. Charact. 129 (2017) 24-30. https://doi.org/10.1016/j.matchar.2017.04.001

[95] N. Wang, Y. Cai, R.Q. Zhang, Growth of nanowires, Mater. Sci. Eng. R Reports. 60 (2008) 1-51. https://doi.org/10.1016/j.mser.2008.01.001

[96] V. Schmidt, J. V. Wittemann, S. Senz, U. Gösele, Silicon Nanowires: A Review on aspects of their growth and their electrical properties, Adv. Mater. 21 (2009) 2681-2702. https://doi.org/10.1002/adma.200803754

[97] N. Fukata, Impurity doping in silicon nanowires, Adv. Mater. 21 (2009) 2829-2832. https://doi.org/10.1002/adma.200900376

[98] J. Shi, X. Wang, Functional semiconductor nanowires via vapor deposition, J. Vac. Sci. Technol. B, Nanotechnol. Microelectron. Mater. Process. Meas. Phenom. 29 (2011) 060801. https://doi.org/10.1116/1.3641913

[99] H. Schift, Nanoimprint lithography: An old story in modern times A review, J. Vac. Sci. Technol. B Microelectron. Nanom. Struct. 26 (2008) 458. https://doi.org/10.1116/1.2890972

[100] K. Peng, Y. Yan, S. Gao, J. Zhu, Dendrite-assisted growth of silicon nanowires in electroless metal deposition, Adv Funct. Mater. 13 (2003) 127-132. https://doi.org/10.1002/adfm.200390018

[101] Z. Huang, N. Geyer, P. Werner, J. De Boor, U. Gösele, Metal-assisted chemical etching of silicon: A review, Adv. Mater. 23 (2011) 285. https://doi.org/10.1002/adma.201001784

[102] W. McSweeney, H. Geaney, C. O'Dwyer, Metal-assisted chemical etching of silicon and the behavior of nanoscale silicon materials as Li-ion battery anodes, Nano Res. 8 (2015) 1395-1442. https://doi.org/10.1007/s12274-014-0659-9

[103] A. Stafiniak, J. Prażmowska, W. Macherzyński, R. Paszkiewicz, Nanostructuring of Si substrates by a metal-assisted chemical etching and dewetting process, RSC Adv. 8 (2018) 31224-31230. https://doi.org/10.1039/C8RA03711F

[104] Q. Wee, J.-W. Ho, S.-J. Chua, Optimized silicon nanostructures formed by one-step metal-assisted chemical etching of Si(111) wafers for GaN deposition, ECS J. Solid State Sci. Technol. 3 (2014) P192-P197. https://doi.org/10.1149/2.009406jss

[105] W.F. Cai, K.B. Pu, Q. Ma, Y.H. Wang, Insight into the fabrication and perspective of dendritic ag nanostructures, J. Exp. Nanosci. 12 (2017) 319-337. https://doi.org/10.1080/17458080.2017.1335890

[106] K. Rajkumar, R. Pandian, A. Sankarakumar, R.T. Rajendra Kumar, Engineering silicon to porous silicon and silicon nanowires by metal-assisted chemical etching: Role of Ag size and electron-scavenging rate on morphology control and mechanism, ACS Omega. 2 (2017) 4540-4547. https://doi.org/10.1021/acsomega.7b00584

[107] J.M. Weisse, C.H. Lee, D.R. Kim, L. Cai, P.M. Rao, X. Zheng, Electroassisted transfer of vertical silicon wire arrays using a sacrificial porous silicon layer, Nano Lett. 13 (2013) 4362-4368. https://doi.org/10.1021/nl4021705

[108] Z. Huang, T. Shimizu, S. Senz, Z. Zhang, X. Zhang, W. Lee, N. Geyer, U. Gösele, Ordered arrays of vertically aligned [110] silicon nanowires by suppressing the crystallographically preferred <100> etching directions, Nano Lett. 9 (2009) 2519-2525. https://doi.org/10.1021/nl303558n

[109] S.-W. Chang, V.P. Chuang, S.T. Boles, C.A. Ross, C. V. Thompson, Densely packed arrays of ultra-high-aspect-ratio silicon nanowires fabricated using block-copolymer lithography and metal-assisted etching, Adv. Funct. Mater. 19 (2009) 2495-2500. https://doi.org/10.1002/adfm.200900181

[110] K.Q. Peng, J.J. Hu, Y.J. Yan, Y. Wu, H. Fang, Y. Xu, S.T. Lee, J. Zhu, Fabrication of single-crystalline silicon nanowires by scratching a silicon surface with catalytic metal particles, Adv. Funct. Mater. 16 (2006) 387-394. https://doi.org/10.1002/adfm.200500392

[111] Y. Harada, X. Li, P.W. Bohn, R.G. Nuzzo, Catalytic amplification of the soft lithographic patterning of Si. nonelectrochemical orthogonal fabrication of photoluminescent porous Si pixel arrays, J. Am. Chem. Soc. 123 (2001) 8709-8717. https://doi.org/10.1021/ja010367j

[112] L. Ji, H. Zheng, A. Ismach, Z. Tan, S. Xun, E. Lin, V. Battaglia, V. Srinivasan, Y. Zhang, Graphene/Si multilayer structure anodes for advanced half and full lithium-ion cells, Nano Energy. 1 (2012) 164-171. https://doi.org/10.1016/j.nanoen.2011.08.003

[113] A. Magasinski, P. Dixon, B. Hertzberg, A. Kvit, J. Ayala, G. Yushin, High-performance lithium-ion anodes using a hierarchical bottom-up approach, Nat. Mater. 9 (2010) 353-358. https://doi.org/10.1038/nmat2725

[114] C.-Y. Wu, C.-C. Chang, J.-G. Duh, Silicon nitride coated silicon thin film on three dimensions current collector for lithium ion battery anode, J. Power Sources. 325 (2016) 64-70. https://doi.org/10.1016/j.jpowsour.2016.06.025

[115] Y. Fan, K. Huang, Q. Zhang, Q. Xiao, X. Wang, X. Chen, Novel silicon-nickel cone arrays for high performance LIB anodes, J. Mater. Chem. 22 (2012) 20870. https://doi.org/10.1039/c2jm34337a

[116] Z. Lu, J. Zhu, D. Sim, W. Zhou, W. Shi, H.H. Hng, Q. Yan, Synthesis of ultrathin silicon nanosheets by using graphene oxide as template, Chem. Mater. 23 (2011) 5293-5295. https://doi.org/10.1021/cm202891p

[117] W.-S. Kim, Y. Hwa, J.-H. Shin, M. Yang, H.-J. Sohn, S.-H. Hong, Scalable synthesis of silicon nanosheets from sand as an anode for Li-ion batteries, Nanoscale. 6 (2014) 4297. https://doi.org/10.1039/c3nr05354g

[118] J. Ryu, D. Hong, S. Choi, S. Park, Synthesis of Ultrathin Si Nanosheets from natural clays for lithium-ion battery anodes, ACS Nano. 10 (2016) 2843-2851. https://doi.org/10.1021/acsnano.5b07977

[119] T.H. Hwang, Y.M. Lee, B.-S. Kong, J.-S. Seo, J.W. Choi, Electrospun core-shell fibers for robust silicon nanoparticle-based lithium ion battery anodes, Nano Lett. 12 (2012) 802-807. https://doi.org/10.1021/nl203817r

[120] P.P. Wang, Y.X. Zhang, X.Y. Fan, J.X. Zhong, K. Huang, Synthesis of Si nanosheets by using sodium chloride as template for high-performance lithium-ion battery anode material, J. Power Sources. 379 (2018) 20-25. https://doi.org/10.1016/j.jpowsour.2018.01.030

[121] S.-H. Ng, J. Wang, D. Wexler, K. Konstantinov, Z.-P. Guo, H.-K. Liu, Highly reversible lithium storage in spheroidal carbon-coated silicon nanocomposites as anodes for lithium-ion batteries, Angew. Chemie Int. Ed. 45 (2006) 6896-6899. https://doi.org/10.1002/anie.200601676

[122] H. Kim, J. Cho, Superior lithium electroactive mesoporous Si@Carbon core−shell nanowires for lithium battery anode material, Nano Lett. 8 (2008) 3688-3691. https://doi.org/10.1021/nl801853x

[123] H. Wu, G. Yu, L. Pan, N. Liu, M.T. McDowell, Z. Bao, Y. Cui, Stable Li-ion battery anodes by in-situ polymerization of conducting hydrogel to conformally coat silicon nanoparticles, Nat. Commun. 4 (2013) 1943. https://doi.org/10.1038/ncomms2941

[124] S.H. Ng, J. Wang, D. Wexler, S.Y. Chew, H.K. Liu, Amorphous Carbon-Coated Silicon Nanocomposites: A low-temperature synthesis via spray pyrolysis and their application as high-capacity anodes for lithium-ion batteries, J. Phys. Chem. C. 111 (2007) 11131-11138. https://doi.org/10.1021/jp072778d

[125] N. Liu, Z. Lu, J. Zhao, M.T. McDowell, H.-W. Lee, W. Zhao, Y. Cui, A pomegranate-inspired nanoscale design for large-volume-change lithium battery anodes, Nat. Nanotechnol. 9 (2014) 187-192. https://doi.org/10.1038/nnano.2014.6

[126] Y. Li, K. Yan, H.-W. Lee, Z. Lu, N. Liu, Y. Cui, Growth of conformal graphene cages on micrometre-sized silicon particles as stable battery anodes, Nat. Energy. 1 (2016) 15029. https://doi.org/10.1038/nenergy.2016.17

[127] X. Xia, J. Tu, Y. Zhang, X. Wang, C. Gu, X.B. Zhao, H.J. Fan, High-quality metal oxide core/shell nanowire arrays on conductive substrates for electrochemical energy storage, ACS Nano. 6 (2012) 5531. https://doi.org/10.1021/nn301454q

[128] J. Wang, Q. Zhang, X. Li, B. Zhang, L. Mai, K. Zhang, Smart construction of three-dimensional hierarchical tubular transition metal oxide core/shell heterostructures with high-capacity and long-cycle-life lithium storage, Nano Energy. 12 (2015) 437-446. https://doi.org/10.1016/j.nanoen.2015.01.003

[129] C. Zhang, T.H. Kang, J.S. Yu, Three-dimensional spongy nanographene-functionalized silicon anodes for lithium ion batteries with superior cycling stability, Nano Res. 11 (2018) 233-245. https://doi.org/10.1007/s12274-017-1624-1

[130]D. Su, M. Cortie, G. Wang, Fabrication of N-doped graphene-carbon nanotube hybrids from Prussian Blue for lithium-sulfur batteries, Adv. Energy Mater. 7 (2017) 1602014. https://doi.org/10.1002/aenm.201602014

[131]L.-F. Chen, S.-X. Ma, S. Lu, Y. Feng, J. Zhang, S. Xin, S.-H. Yu, Biotemplated synthesis of three-dimensional porous MnO/C-N nanocomposites from renewable rapeseed pollen: An anode material for lithium-ion batteries, Nano Res. 10 (2017) 1-11. https://doi.org/10.1007/s12274-016-1283-7

[132]N. Geyer, B. Fuhrmann, H.S. Leipner, P. Werner, Ag-mediated charge transport during metal-assisted chemical etching of silicon nanowires, ACS Appl. Mater. Interfaces. 5 (2013) 4302-4308. https://doi.org/10.1021/am400510f

[133] H.D. Um, N. Kim, K. Lee, I. Hwang, J. Hoon Seo, Y.J. Yu, P. Duane, M. Wober, K. Seo, Versatile control of metal-assisted chemical etching for vertical silicon microwire arrays and their photovoltaic applications, Sci. Rep. 5 (2015) 11277. https://doi.org/10.1038/srep11277

[134]A.T. Tesfaye, R. Gonzalez-Rodriguez, J.L. Coffer, T. Djenizian, Self-supported silicon nanotube arrays as an anode electrode for Li-ion batteries, ECS Trans. 77 (2017) 349-350. https://doi.org/10.1149/07711.0349ecst

[135]O. Pérez-Díaz, E. Quiroga-González, N.R. Silva-González, Silicon microstructures through the production of silicon nanowires by metal-assisted chemical etching, used as sacrificial material, J. Mater. Sci. 54 (2019) 2351-2357. https://doi.org/10.1007/s10853-018-3003-z

[136]X. Chen, Q. Bi, M. Sajjad, X. Wang, Y. Ren, X. Zhou, W. Xu, Z. Liu, One-dimensional porous silicon nanowires with large surface area for fast charge-discharge lithium-ion batteries, Nanomaterials. 8 (2018) 285. https://doi.org/10.3390/nano8050285

[137] F. Sun, K. Huang, X. Qi, T. Gao, Y. Liu, X. Zou, X. Wei, J. Zhong, A rationally designed composite of alternating strata of Si nanoparticles and graphene: A high-performance lithium-ion battery anode, Nanoscale. 5 (2013) 8586. https://doi.org/10.1039/c3nr02435k

[138] S. Karthik Kiran, S. Shukla, A. Struck, S. Saxena, Surface engineering of graphene oxide shells using lamellar LDH nanostructures, ACS Appl. Mater. Interfaces. 11 (2019) 20232-20240. https://doi.org/10.1021/acsami.8b21265

[139]T. Wang, J. Zhu, Y. Chen, H. Yang, Y. Qin, F. Li, Q. Cheng, X. Yu, Z. Xu, B. Lu, Large-scale production of silicon nanoparticles@graphene embedded in nanotubes as ultra-robust battery anodes, J. Mater. Chem. A. 5 (2017) 4809-4817. https://doi.org/10.1039/C6TA10631E

[140] S. Karthik Kiran, S. Shukla, A. Struck, S. Saxena, Surface enhanced 3D rGO hybrids and porous rGO nano-networks as high performance supercapacitor electrodes for integrated energy storage devices, Carbon 158 (2019) 527-535. https://doi.org/10.1016/j.carbon.2019.11.021

[141] Z. Lu, N. Liu, H.-W. Lee, J. Zhao, W. Li, Y. Li, Y. Cui, Nonfilling carbon coating of porous silicon micrometer-sized particles for high-performance lithium battery anodes, ACS Nano. 9 (2015) 2540-2547. https://doi.org/10.1021/nn505410q

[142] X. Tang, G. Wen, Y. Zhang, D. Wang, Y. Song, Novel silicon nanoparticles with nitrogen-doped carbon shell dispersed in nitrogen-doped graphene and CNTs hybrid electrode for lithium ion battery. Appl. Surf. Sci. 425 (2017) 742-749. https://doi.org/10.1016/j.apsusc.2017.07.058

[143] K.G. Gallagher, S.E. Trask, C. Bauer, T. Woehrle, S.F. Lux, M. Tschech, P. Lamp, B.J. Polzin, S. Ha, B. Long, Q. Wu, W. Lu, D.W. Dees, A.N. Jansen, Optimizing areal capacities through understanding the limitations of lithium-ion electrodes, J. Electrochem. Soc. 163 (2016) A138-A149. https://doi.org/10.1149/2.0321602jes

[144] N. Lin, Y. Han, L. Wang, J. Zhou, J. Zhou, Y. Zhu, Y. Qian, Preparation of nanocrystalline silicon from SiCl4 at 200 °C in molten salt for high-performance anodes for lithium ion batteries, Angew. Chemie Int. Ed. 54 (2015) 3822-3825. https://doi.org/10.1002/anie.201411830

[145] J.-K. Yoo, J. Kim, H. Lee, J. Choi, M.-J. Choi, D.M. Sim, Y.S. Jung, K. Kang, Porous silicon nanowires for lithium rechargeable batteries, Nanotechnology. 24 (2013) 424008. https://doi.org/10.1088/0957-4484/24/42/424008

[146] B. Gattu, R. Epur, P.H. Jampani, R. Kuruba, M.K. Datta, P.N. Kumta, Silicon-carbon core-shell hollow nanotubular configuration high-performance lithium-ion anodes, J. Phys. Chem. C. 121 (2017) 9662-9671. https://doi.org/10.1021/acs.jpcc.7b00057

[147] D. Jia, X. Li, J. Huang, Bio-inspired sandwich-structured carbon/silicon/titanium-oxide nanofibers composite as an anode material for lithium-ion batteries, Compos. Part A Appl. Sci. Manuf. 101 (2017) 273-282. https://doi.org/10.1016/j.compositesa.2017.06.028

[148] S. Chen, Z. Chen, X. Xu, C. Cao, M. Xia, Y. Luo, Scalable 2D mesoporous silicon nanosheets for high-performance lithium-ion battery anode, Small. 14 (2018) 1703361. https://doi.org/10.1002/smll.201703361

[149] L. Yan, J. Liu, Q. Wang, M. Sun, Z. Jiang, C. Liang, F. Pan, Z. Lin, In Situ wrapping Si nanoparticles with 2D carbon nanosheets as high-areal-capacity anode for

lithium-ion batteries, ACS Appl. Mater. Interfaces. 9 (2017) 38159-38164.
https://doi.org/10.1021/acsami.7b10873

[150] P. Gao, H. Tang, A. Xing, Z. Bao, Porous silicon from the magnesiothermic
reaction as a high-performance anode material for lithium ion battery applications,
Electrochim. Acta. 228 (2017) 545-552.
https://doi.org/10.1016/j.electacta.2017.01.119

[151] K. Zhang, Y. Xia, Z. Yang, R. Fu, C. Shen, Z. Liu, Structure-preserved 3D porous
silicon/reduced graphene oxide materials as anodes for Li-ion batteries, RSC Adv. 7
(2017) 24305-24311. https://doi.org/10.1039/C7RA02240A

Keyword Index

About the Editors

Dr. Inamuddin is currently working as Assistant Professor in the Chemistry Department, Faculty of Science, King Abdulaziz University, Jeddah, Saudi Arabia. He is a permanent faculty member (Assistant Professor) at the Department of Applied Chemistry, Aligarh Muslim University, Aligarh, India. He obtained Master of Science degree in Organic Chemistry from Chaudhary Charan Singh (CCS) University, Meerut, India, in 2002. He received his Master of Philosophy and Doctor of Philosophy degrees in Applied Chemistry from Aligarh Muslim University (AMU), India, in 2004 and 2007, respectively. He has extensive research experience in multidisciplinary fields of Analytical Chemistry, Materials Chemistry, and Electrochemistry and, more specifically, Renewable Energy and Environment. He has worked on different research projects as project fellow and senior research fellow funded by University Grants Commission (UGC), Government of India, and Council of Scientific and Industrial Research (CSIR), Government of India. He has received Fast Track Young Scientist Award from the Department of Science and Technology, India, to work in the area of bending actuators and artificial muscles. He has completed four major research projects sanctioned by University Grant Commission, Department of Science and Technology, Council of Scientific and Industrial Research, and Council of Science and Technology, India. He has published 171 research articles in international journals of repute and eighteen book chapters in knowledge-based book editions published by renowned international publishers. He has published 105 edited books with Springer (U.K.), Elsevier, Nova Science Publishers, Inc. (U.S.A.), CRC Press Taylor & Francis Asia Pacific, Trans Tech Publications Ltd. (Switzerland), IntechOpen Limited (U.K.), Wiley-Scrivener, (U.S.A.) and Materials Research Forum LLC (U.S.A). He is a member of various journals' editorial boards. He is also serving as Associate Editor for journals (Environmental Chemistry Letter, Applied Water Science and Euro-Mediterranean Journal for Environmental Integration, Springer-Nature), Frontiers Section Editor (Current Analytical Chemistry, Bentham Science Publishers), Editorial Board Member (Scientific Reports-Nature), Editor (Eurasian Journal of Analytical Chemistry), and Review Editor (Frontiers in Chemistry, Frontiers, U.K.) He is also guest-editing various special thematic special issues to the journals of Elsevier, Bentham Science Publishers, and John Wiley & Sons, Inc. He has attended as well as chaired sessions in various international and national conferences. He has worked as a Postdoctoral Fellow, leading a research team at the Creative Research Initiative Center for Bio-Artificial Muscle, Hanyang University, South Korea, in the field of renewable energy, especially biofuel cells. He has also worked as a Postdoctoral Fellow at the Center of Research Excellence in Renewable Energy, King Fahd University of Petroleum and Minerals, Saudi Arabia, in the field of polymer electrolyte membrane fuel cells and computational fluid dynamics of polymer electrolyte membrane fuel cells. He is a life member of the Journal of the Indian Chemical Society. His research interest includes ion exchange materials, a sensor for heavy metal ions, biofuel cells, supercapacitors and bending actuators.

Dr. Rajender Boddula is currently working with Chinese Academy of Sciences-President's International Fellowship Initiative (CAS-PIFI) at National Center for Nanoscience and Technology (NCNST, Beijing). He obtained Master of Science in Organic Chemistry from Kakatiya University, Warangal, India, in 2008. He received his Doctor of Philosophy in Chemistry with the highest honours in 2014 for the work entitled "Synthesis and Characterization of Polyanilines for Supercapacitor and Catalytic Applications" at the CSIR-Indian Institute of Chemical Technology (CSIR-IICT) and Kakatiya University (India). Before joining National Center for Nanoscience and Technology (NCNST) as CAS-PIFI research fellow, China, worked as senior research associate and Postdoc at National Tsing-Hua University (NTHU, Taiwan) respectively in the fields of bio-fuel and CO_2 reduction applications. His academic honors include University Grants Commission National Fellowship and many merit scholarships, study-abroad fellowships from Australian Endeavour Research Fellowship, and CAS-PIFI. He has published many scientific articles in international peer-reviewed journals and has authored around twenty book chapters, and he is also serving as an editorial board member and a referee for reputed international peer-reviewed journals. He has published edited books with Springer (UK), Elsevier, Materials Research Forum LLC (USA), Wiley-Scrivener, (U.S.A.) and CRC Press Taylor & Francis group. His specialized areas of research are energy conversion and storage, which include sustainable nanomaterials, graphene, polymer composites, heterogeneous catalysis for organic transformations, environmental remediation technologies, photoelectrochemical water-splitting devices, biofuel cells, batteries and supercapacitors.

Dr. Mohammad Faraz Ahmer is presently working as Assistant Professor in the Department of Electrical Engineering, Mewat Engineering College, Nuh Haryana, India, since 2012 after working as Guest Faculty in University Polytechnic, Aligarh Muslim University Aligarh, India, during 2009-2011. He completed M.Tech. (2009) and Bachelor of Engineering (2007) degrees in Electrical Engineering from Aligarh Muslim University, Aligarh in the first division. He obtained a Ph.D. degree in 2016 on his thesis entitled "Studies on Electrochemical Capacitor Electrodes". He has published six research papers in reputed scientific journals. He has edited two books with Materials Research Forum, U.S.A. His scientific interests include electrospun nano-composites and supercapacitors. He has presented his work at several conferences. He is actively engaged in searching of new methodologies involving the development of organic composite materials for energy storage systems.

Prof. Abdullah M. Asiri is the Head of the Chemistry Department at King Abdulaziz University since October 2009 and he is the founder and the Director of the Center of Excellence for Advanced Materials Research (CEAMR) since 2010 till date. He is the Professor of Organic Photochemistry. He graduated from King Abdulaziz University (KAU) with B.Sc. in Chemistry in 1990 and a Ph.D. from University of Wales, College of Cardiff, U.K. in 1995. His research interest covers color chemistry, synthesis of novel photochromic and thermochromic systems, synthesis of novel coloring matters and dyeing of textiles, materials chemistry, nanochemistry and nanotechnology, polymers and plastics. Prof. Asiri is the principal supervisors of more than 20 M.Sc. and six Ph.D. theses. He is the main author of ten books of different chemistry disciplines. Prof. Asiri is the Editor-in-Chief of King Abdulaziz University Journal of Science. A major achievement of Prof. Asiri is the research of tribochromic compounds, a new class of compounds which change from slightly or colorless to deep colored when subjected to small pressure or when grind. This discovery was introduced to the scientific community as a new terminology published by International Union of Pure and Applied Chemistry (IUPAC) in 2000. This discovery was awarded a patent from European Patent office and from UK patent. Prof. Asiri involved in many committees at the KAU level and on the national level. He took a major role in the advanced materials committee working for King Abdulaziz City for Science and Technology (KACST) to identify the national plan for science and technology in 2007. Prof. Asiri played a major role in advancing the chemistry education and research in KAU. He has been awarded the best researchers from KAU for the past five years. He also awarded the Young Scientist Award from the Saudi Chemical Society in 2009 and also the first prize for the distinction in science from the Saudi Chemical Society in 2012. He also received a recognition certificate from the American Chemical Society (Gulf region Chapter) for the advancement of chemical science in the Kingdome. He received a Scopus certificate for the most publishing scientist in Saudi Arabia in chemistry in 2008. He is also a member of the editorial board of various journals of international repute. He is the Vice- President of Saudi Chemical Society (Western Province Branch). He holds four USA patents, more than one thousand publications in international journals, several book chapters and edited books.